Dear David,

I just want to tell you how much I enjoyed your book "The Long Return"!

I had three tours in Vietnam, 61, 63, and 67 -- flying T-28s and B-26s out of Bien Hoa the first two and F-4s out of Ubon the last one!

Sincerely,
"Pete" Piotrowski

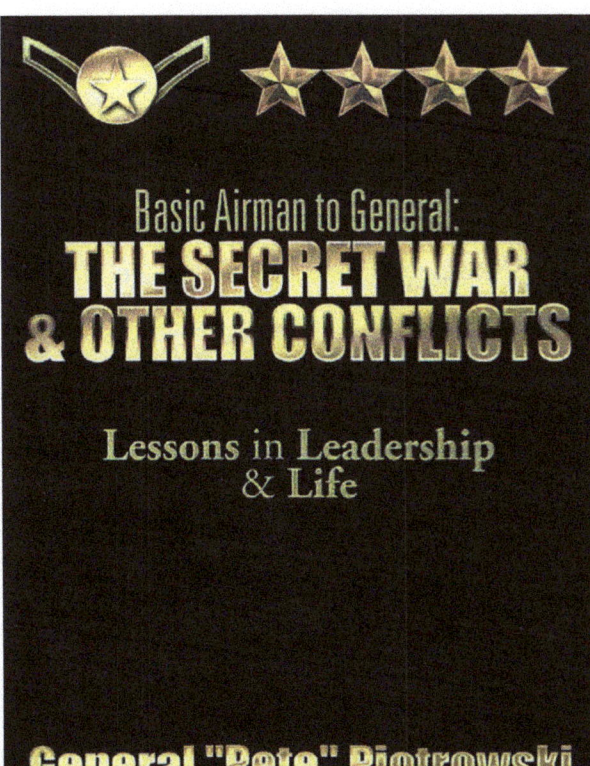

Basic Airman to General:
THE SECRET WAR & OTHER CONFLICTS

Lessons in Leadership & Life

General "Pete" Piotrowski

978-1-965552-29-2 (Paperback)

BOOKWRIGHTS
HOUSE

admin@bookwrightshouse.com
☎ (213) 286 6700

THE LONG RETURN

Should the Military Be Used as a Political Tool?

Col. David O. Scheiding, USAF, (RET)

David O. Scheiding is also the author of
Hank: An "Angel Dog,"
and
"Don't Fly Today"

SHOULD THE MILITARY BE USED AS A POLITICAL TOOL?

ACKNOWLEDGMENTS

I WOULD LIKE TO thank the following people for all of their help with this book:

I want to thank my wife of sixty-two years for putting up with me and helping me, her dumb fighter pilot and engineer husband, to not only write one book but three. I am also truly indebted to Ms. Mary Jo Salas, who word processed the manuscript on her own time. I also need to thank my son, Doug Scheiding, for helping me put all of the old photos into a digital format to enhance the quality of the old 1970s photos. Even though I do not know him, I have to thank 1st LT. Lee Alley for his book, titled *Back from War*. Without it, I still may not have mentally returned home.

This book is dedicated to the 238 Forward Air Controllers (FACs) who did not return home from Vietnam.

CONTENTS

FOREWORD

THE VIETNAM WAR WAS a very costly war for the United States (U.S.) in both resources and loss of American lives. The total U.S. casualties during this conflict exceeded 58,000 lives, and the cost exceeded 150 billion dollars. The conflict lasted for approximately 10 years. The risks to U.S. ground forces and Air Force pilots were significant. For the Air Force, the risks were greatest for fighter pilots and forward air controllers (FACs).

Fighter pilots who flew F-105s or F-4s over North Vietnam had to fly 100 missions over the North before they could come home. *For these pilots, it was said that there was a 40 percent to 50 percent probability that these pilots would be shot down before completing 100 missions.* The Air Force, however, did not have any problems getting volunteers to fly fighters.

Forward Air Controllers (FACs) flew O-1As, O-2As or OV-10s. The O-2A was essentially a light aircraft (a Cessna Skymaster) with two engines. The O-2A took off at 90 knots, climbed-out at 90 knots, landed at 90 knots and cruised at 120 knots. The O-1A was similar but was a single engine light aircraft that flew a little slower. OV-10s were faster but still were propeller driven. The OV-10s did, however, have mini-guns to help defend themselves somewhat.

The O-1As and O-2As flew at altitudes mostly between 1500 and 2500 feet during their combat missions. This placed them constantly in the small arms envelope for engagement. OV-10s could fly higher and faster but still were much slower than fighter aircraft.

The Air Force initially tried to use O-2A aircraft over the Ho Chi Minh Trail network in Laos. The survival rate for O-2As over the trails was similar to the F-105 survival rate over North Vietnam. The Air Force later only allowed OV-10s to operate over the trail system in Laos where they had a much better survival rate.

During the 10 years of the Vietnam War, the united States military lost a total of 9929 aircraft. Of this total, 3744 were fixed wing aircraft. Of these 3744 fixed wing aircraft loses, the Air Force lost a total of 1489 fighter and Forward Air Controller (FAC) aircraft to combat losses. This works out to be a percentage of 39.7 probability of being shot down as either a fighter pilot or FAC before completing their tour. Being a FAC or fighter pilot was a very dangerous business.

During the 10 years of FAC operations, the Air Force lost a total of 238 FACs. That is almost two FACs a month for 10 years. Being a FAC was very risky but very rewarding. Of the 12 Medal of Honor recipients of the Vietnam War, two were FACs. This is the story of one FAC's tour of duty in Vietnam during 1971 and 1972 and the difficulties he experienced upon returning to the United States.

PROLOGUE

EVEN THOUGH I PHYSICALLY returned to the United States the first time I came home from Vietnam in late May 1972, I was one of those who actually did not come home until much, much later. Upon my return in 1972, the United States that I remembered from the 1968 to 1970 time period had changed significantly. I was met with what I would call total culture shock. This was based on the amount of change that I noticed in the general American attitude toward the war in Vietnam and those of us who served there. I was met with a tremendous amount of anti-war and anti-military sentiment by the mainstream liberal news media and from society in general. This totally shocked me. This resulted in confusion for me as I tried to re-assimilate back into this new and different environment from the one that I had left.

After I physically returned from Vietnam in 1972, my wife felt I had changed. I did not think that I had, but I did not want to talk about Vietnam. I was trying to process the current environment that I now was in. It was so different from the one that I had left. I had to figure out just how I now fitted into this new environment. I am sure that I was not the easiest person to live with during that time period.

On March 6, 2009, my wife came across a book titled *Back from War* by 1st Lt. Lee Alley. She bought the book and gave it to me to read. The subtitle of the book is *Finding Hope & Understanding in Life after Combat for the American Solider & Their Loved Ones.*

This book contains stories written by different soldiers about their experiences upon returning to the States from Vietnam. While reading this book, I saw myself in many of the stories and could relate to these individuals who had expressed exactly how I felt upon my return. In one story, an individual had done the same thing with his medals that I had done with mine. Reading this book, I realized that I was not alone. I found out that there were many soldiers going through the same thing that I was going through.

After I finished reading this book, I was finally able to return home mentally. For the first time I was able to talk about Vietnam with my family and others. I finally completed my two returns to the United States from Vietnam in April 2009.

INTRODUCTION

WHEN I LEFT FOR Vietnam the first time in 1971, my youngest son had not even been born; and my oldest son was four years old. My oldest son grew up and attended Texas A&M University and graduated in 1989. My wife and I had been buying Texas A&M season football tickets since 1986. It was after April 2009, while on some of these trips to College Station and back to San Antonio to attend football games, I began to relate some of my Vietnam experiences to my wife. After I had related a number of stories to her, she suggested that it would be nice if I could provide my sons with some kind of record about my experiences in Vietnam. I had not wanted to talk about Vietnam until after I had completed reading the book my wife had bought for me in 2009. I thought that this was a very good idea as I would have liked to have had talked to my dad about Vietnam before he passed away in 1991.

In late 2009, I prepared a cassette tape for my sons that was approximately one hour and 15 minutes long about some of my Vietnam experiences. Since then technology has progressed to the point that I doubt that either of my sons would have or could find a functioning cassette tape player to even listen to the tape.

Because of this, in 2015, I decided to write this manuscript to document my experiences. I also now have my medals from Vietnam displayed on a wall in a shadow box that my oldest son had given me for Christmas many years before. In addition, in 2009, I had the first individual thank me for my military service. I can now say that I am truly home and at peace with myself.

This is the story of my long return from Vietnam and why. It also documents the difficulties that I have had justifying the fact that our government leaders use the military as a political tool.

CHAPTER ONE

The Long Departure

I ENTERED THE UNITED States Air Force (USAF) in December 1964 upon graduation from Iowa State University (ISU) through the Reserve Officer Training and Commissioning (ROTC) program. At that time, ROTC was mandatory for the first two years of college at State Universities. I decided to sign up for advanced ROTC (the last two years) if I could pass the flight physical for pilot training. I did pass the flight physical, and after completing the last two years of ROTC, I graduated with a Bachelor of Science (BS) degree in Aerospace Engineering and was commissioned as a 2nd Lieutenant in the United States Air Force (USAF).

My first assignment was Undergraduate Pilot Training (UPT) at Reese Air Force Base (AFB) in Lubbock, Texas. After completing UPT in 1966, I was assigned as a T-37 Instructor Pilot (IP) at Laughlin AFB in Del Rio, Texas. At that time of my graduation from UPT, the Vietnam War was beginning to escalate. There were 37 students in my pilot training class. Of these, nine were assigned to F-4Cs stationed in Europe. They were, however, assigned to the back seat of the F-4 and not to the front seat as pilot in command. At that time, the Air Force was assigning two pilots to F-4s. They later began to assign navigators or weapon systems officers (WSOs) to the back seat of F-4s. To get the front seat as a pilot, one had to volunteer to go to Vietnam at which time one could upgrade to the front seat. In addition, seven of my classmates were assigned to the Tactical Air Command (TAC) or to the Pacific Air Force (PACAF), which were both in direct support to Vietnam. We also had four of my classmates go to the Air Defense Command (ADC), but soon they were on their way to Vietnam. I knew that if I remained in the Air Force, I would eventually be going to Vietnam.

While at Laughlin AFB in 1968, the Air Force initiated a program to entice pilots to volunteer to go to Vietnam. The program consisted of being able to volunteer for the type of aircraft system one would like to fly in Vietnam. The types of aircraft categories that one could volunteer for consisted of airlift, bombers, or fighter/forward air controller (FAC) type of aircraft. By volunteering, one could at least have a say in the type of aircraft and job one would be doing in Vietnam. This was opposed to just waiting for an assignment to come down from Air Force and taking whatever the requirement was at that time. This program also helped the Air Force to say that all of the pilots going to Vietnam were volunteers.

In 1968 I volunteered to go to Vietnam in fighter/forward air controller (FAC) type aircraft. I volunteered as I felt it was my duty as an Air Force officer in order to serve my country. I was not selected to go until December 1970. I was told by the Air Force that I was needed more as an Instructor Pilot (IP) to train pilots than for me to go to Vietnam. I knew, however, that eventually I would be selected to go. This was the only assignment pilots at Laughlin AFB, who were staying in the Air Force, were receiving. Many pilots were separating from the Air Force in lieu of going to Vietnam. Since I felt a duty to my country, separating from the Air Force was not an option for me.

Even though I had volunteered for Vietnam in mid-1968, it was not until December 1970 that I finally reached the top of the volunteer list at Laughlin AFB. The assignment I received was as an O-2A Forward Air Controller (FAC) with Fighter Qualification. I was a little disappointed to be assigned as a FAC because I really wanted to fly fighters; however, FAC was part of the fighter group of aircraft; and I did at least get a FAC assignment with fighter qualification required. The Army had a requirement that any FAC assigned to support them had to be fighter qualified. This meant that I would most likely be assigned to support the Army while in Vietnam.

I left Laughlin AFB in January 1971 for six months of training before my departure for Vietnam, which was scheduled for June 1971. My training consisted of two months of fighter qualification training at Cannon AFB in New Mexico, followed by water survival training at Homestead AFB in Florida and basic survival training at Fairchild AFB in Washington State. The final month and a half of training was O-2A aircraft training at Hurlburt Field at Fort Walton Beach, Florida.

During the fighter qualification training at Cannon AFB, we received weekly briefings on how the war was going and number of Air Force casualties. Since I had spent almost two and a half years on the volunteer list, a number of pilots from Laughlin AFB had departed before me. At the end of the briefings they would list the names of the Air Force casualties from the week before. There were always four to six names on these lists. It seemed that I recognized at least one name on each of the lists during each briefing. Laughlin AFB was not doing very well on getting pilots back from Vietnam. In fact, the first 19 pilots selected off of the volunteer list at Laughlin AFB did not return from Vietnam. They were either listed as killed in action (KIA) or shot down over North Vietnam and became prisoners of war (POWs). It was just before I left Laughlin AFB in January 1971, that Laughlin finally got three pilots back from Vietnam. One was even an F-105 pilot who had flown his 100 missions over North Vietnam. There finally was a ray of hope for us volunteers going to Vietnam from Laughlin AFB.

I left the States in late June 1971 for jungle survival school at Clark AB in the Philippines. I arrived in Vietnam in July 1971 for my tour of duty as an O-2A Forward Air Controller (FAC). My departure for Vietnam had lasted from mid-1968, when I volunteered, to my arrival in Vietnam in July 1971 approximately three years later.

CHAPTER TWO

The Arrival

IT WAS JULY 1971 when I finally arrived in Vietnam to begin my year-long tour as an O-2A Forward Air Controller (FAC). I arrived on a military charter Continental flight that landed at Cam Ranh Bay in Vietnam's Military Region II (MRII). After we landed, the aircraft did not taxi up to the terminal to deplane but was escorted and parked away from the terminal in what was known as the "hot brake" area. We were directed to deplane and wait for buses to arrive to transport the 200-plus soldiers from all branches of service that were on board to the terminal.

As we waited for the buses, the aircrew that had been flying the aircraft had gathered under the tail section and were looking up at the horizontal stabilizer. Being a pilot myself, I walked over to the crew to see what they were looking at, knowing that something was amiss, since we had not been allowed to taxi up to the terminal for deplaning. As we looked up at the horizontal stabilizer, we saw an obvious bullet hole. The aircraft had taken a 50-caliber round through the horizontal stabilizer on final approach and landing. That was the reason we had been directed to park in the "hot brake" area. As I looked up at the bullet hole, I thought to myself that this is going to be a very long year.

As a FAC, I was assigned to the 504th Tactical Air Support Group (TASG), which was located at Cam Ranh Bay. South Vietnam had been divided up into four Military Regions for combat responsibility operations (See Figure 1). The 504th Group had three Tactical Air Support Squadrons (TASSs) under its control. The three Tactical Air Support Squadrons (TASSs) consisted of the 19th TASS, the 20th TASS and the 21st TASS. There was one squadron assigned to each of the three Military Regions (MRs) for which the Air Force had responsibility for tactical air support in South Vietnam. These MRs were MRI, MRII and MRIII. The Navy had tactical air support responsibility for MRIV. These Military Regions were also referred to as I Corps, II Corps, III Corps and IV Corps.

The 504th TASG and each of its three TASSs are not like the normal Air Force Tactical Fighter Wing (TFW) and Tactical Fighter Squadrons (TFSs). In a normal Tactical Fighter Wing (TFW), which is located at one main base of operations, there are three or four Tactical Fighter Squadrons (TFSs) assigned to the Wing and are under direct operational control of the Wing. Under the 504th TASG, the three TASSs were not co-located with the 504th TASG nor were they under the direct operational control of the Group. This is why it was called a Group and not a Wing.

Each one of the three TASSs were located at a different main base for operations in each of the three Military Regions and operated with a significant amount of autonomy from the Group. Each TASS had direct operational control of daily operations in their Areas of Responsibility (AR) for combat tactical air support. In addition to the TASS's main operating base, each TASS also operated out of forward operating locations (FOLs). At each FOL, four or five aircraft were permanently based along with 10 to 12 pilots and a contingent of maintenance personnel to maintain the aircraft. This is also different from TFSs and TFWs in that maintenance personnel are not assigned directly to each of the TFSs. There are maintenance squadrons that are also assigned to the TFW which support each of the flying squadrons.

Each of the FAC FOLs were essentially mini-squadrons that were under the direct control of an Air Liaison Officer (ALO) at each location, usually a Major or Senior Captain. Each FOL also operated independently of the TASS to which they were assigned. Daily operational control over flying operations rested with the ALO at each FOL. The TASS would provide replacement aircraft and personnel when needed at each FOL as required. The responsibility of daily combat flying operations, however, rested solely with the ALO at the FOL. This organizational structure would become a very large potential problem for me during the latter portion of my tour with the drawdown of U.S. forces from South Vietnam.

Being fighter pilot qualified, I was assigned to the 20th Tactical Air Support Squadron (20th TASS) based at Da Nang Air Base in MRI or I Corps (See Figure 2). The 20th TASS had the responsibility of providing tactical air support for the Army's 101st Airborne Division which had the responsibility for combat operations in MRI.

I arrived at Da Nang AB and the 20th TASS three days after arriving in-country. Da Nang is located approximately 100 miles south of the then Demilitarized Zone (DMZ) between North and South Vietnam. The actual border between North and South Vietnam at that time was the 1954 Demarcation Line, which was actually the Ben Hai River. The DMZ consisted of approximately two and a half miles on either side of the river.

After processing into the 20th TASS, I was told that I would be given an in-country orientation checkout in the O-2A aircraft since my last flight had been in May 1971, while in O-2A training in Florida. The checkout was to include both day and night sorties as part of the in-country orientation.

I had my first in-country orientation flight out of Da Nang on July 21, 1971. I received two more day flights out of Da Nang and was told that I had been assigned to Hue Phu Bai forward operating location (FOL) in support of the 2nd Brigade of the 101st Airborne Division. I was also told that I would receive my night in-country orientation checkout flight with the Air Liaison Officer (ALO) in charge at that location—a Major Robbins.

Hue Phu Bai was located approximately 60 miles south of the DMZ and was the main operating base of the 2nd Brigade of the 101st Airborne Division (See Photos 1 and 2). The 2nd Brigade had an Area of Responsibility (AR) from approximately 30 miles south of the DMZ along the east coastline, to 80 miles south and all of the area west to the Laos border. The 1ˢᵗ Brigade of the 101ˢᵗ Airborne Division had the AR from the DMZ south for 30 miles along the eastern coastline and the area west to the Laos border. The FACs that supported the Army's 1ˢᵗ Brigade flew out of Camp Evans (FOL) which was located 20 miles south of the DMZ in Quang Tri Province. Quan Tri Province was not part of the 2ⁿᵈ Brigade's AR that we supported out of the Hue Phu Bai (FOL) for the 20ᵗʰ TASS.

Major Robbins gave me two day rides as part of my local AR checkout and cleared me to fly day air support missions in support of the Army's operations. He said we would do the night checkout later. He called Da Nang (20ᵗʰ TASS) and informed them that my night checkout would be completed later and that he would let them know when it was completed.

CHAPTER THREE

Hue Phu Bai

I WAS NOW AT Phu Bai and ready to settle in and do my duty for my country. It is interesting to note that my living conditions deteriorated as I moved north from Cam Ranh Bay. At Cam Ranh Bay, with the 504th TASG, the living quarters for pilots who had just arrived consisted of small travel trailers that had been divided into two sections. One pilot stayed in each half of the trailer during the time we were at Cam Ranh Bay before being assigned to our end destination. These units did have air conditioning and indoor/outdoor carpeting. The Air Force had a requirement that pilot living quarters had to be air conditioned. Although small, one could get a good night's rest, and it was comfortable and even had an indoor restroom.

At Da Nang Air Base, where the 20th TASS was located, living quarters for pilots were similar to those at Cam Ranh Bay's. We again stayed in these small travel trailer type quarters. Again, they were air conditioned with one pilot to each half of the trailer. At Phu Bai, the living conditions definitely deteriorated. I now knew that I was living with the Army.

The Army helicopter pilots and we FAC pilots were assigned to relatively large wooden hooches (See Photo 3). These hooches were divided up into four separate living quarters. Each hooch had a hallway down the middle with each of the four corners being separated into a living quarters for one pilot. There was a door on each end of the hooch's hallway for entry. One pilot was assigned to each of the four units.

Some of the hooches had air conditioning while the others had only plywood covered screens that when opened provided some air movement along with a fan. At Phu Bai we did not always have electricity to run the air conditioners if they even worked. We normally only had electricity in the living quarters from 1800 hours (6:00 P.M.) to around 2200 hours (10:00 P.M.). During the day, the limited electricity was supplied to the operating facilities like the Tactical Operations Centers (TOCs), the Post Exchange (PX), Mess Halls and Barber Shop. There just was not always enough electricity to go around 24 hours a day.

The latrines consisted of small wooden outdoor structures. The Army would close down a street and place these wooden outhouse latrines along the street (See Photo 4). Fifty-five gallon drums were placed under each of the four holes inside the latrine. South Vietnamese workers would replace these drums each day and take the used ones to a location where the contents would be burned. There was no running water for these latrines as water was also limited. Toilet paper was an individual's responsibility to bring along when using the facility. Toilet paper was like gold, and you had to protect your supply or it would disappear if left out in your living quarters.

As far as water goes, one hooch located in a group of hooches was designated and converted into a central shower facility. We had to walk along wooden plywood pallet sidewalks or, when lucky, on perforated steel planks or panels (PSP) to get to the shower hooch (See Photo 5). We never had hot water, so showering and shaving was completed using cold water. It was not too bad since we were in a jungle environment, and it never was cold. The wooden plywood pallets or PSP kept us from walking in the mud when it rained. Since we did not spend much time in our living quarters, none of these conditions mattered that much.

We did have a bed (See Photo 6), a steel locker where we could lock up our toilet paper (See Photo 7), a fan and a desk to write letters home (See Photo 8). Wall shelves consisted of old rocket boxes or wooden shelves if one took the time to construct them. As crude as it was, it was now home.

In addition, each hooch had a sandbag bunker located outside next to the hooch for the times when rocket or mortar attacks would occur (See Photo 9). Living with the Army was definitely different than living with the Air Force. I did, however, at least have a bed, a steel locker, a fan and a desk, and this was now going to be home for a while.

At Phu Bai we also did not have any potable drinking water available for use. We would have to go to the Post Exchange (PX) and buy cases of sodas to drink (See Photo 10). We would keep these sodas in our living areas, but they were not always cold. Without electricity most of the time in the living quarters, even if we were lucky enough to have a refrigerator, the sodas were warm. About the only good thing about warm sodas is that they are wet.

Even though the Army did have a PX, it was not stocked very well. In addition, when one of the Army units would return from a Fire Base (FB), their first stop was the PX. These individuals would essentially clean out the PX. The day after one of these units returned, the shelves at the PX were essentially bare. About the only thing that was always available were sanitary napkins. There always was a full shelf of these napkins, and they were covered with dust. I never really understood why the Army stocked so many of these napkins. There were very, and I mean very, few females at Phu Bai. I cannot ever remember seeing any American females at Phu Bai.

I had always heard that the Army did not know how to live. I was now experiencing living with the Army.

CHAPTER FOUR

Phu Bai Operations

PHU BAI WAS THE main operating base for the 2nd Brigade of the 101st Airborne Division. Throughout their operating Area of Responsibility (AR), the 2nd Brigade also established forward operating locations called "Fire Bases" (FBs). Units from the 2nd Brigade would be deployed to these FBs for varying lengths of time. From these FBs, combat operations would be conducted around the area of each FB. These FBs each had names and usually were located on top of mountains (See Photos 11 and 12).

The Army would send out reconnaissance patrols from these FBs to identify enemy targets. They would provide this intelligence of possible targets back to Phu Bai for consideration of engagement. The Tactical Operations Center (TOCs) for both the Air Force and the Army were co-located to allow for smoother operations in support of combat activities. This combined center for both the Army and Air Force's TOCs was called the Tactical Air Support Center (TASC). The Air Force FAC operations along with the Army helicopter operations were controlled out of this center. If the Army felt that they could handle the target with their helicopter assets, they would service the target request. If the target was more of a problem, they would request the Air Force to service the target. If a "troops-in-contact" (TIC) situation occurred at any of these FBs, we all would do whatever was necessary to address the situation.

For those targets that were not mobile in nature, we would order air strikes for the next day. We would order these air strikes through the Direct Air Support Center (DASC), which was located at Da Nang AB. The DASC at Da Nang had tactical air support responsibility for all of MRI or I Corps. If tactical air support was needed immediately, we would either launch a FAC or have one of our airborne FACs address the situation with air support. We always had at least two FACs airborne at all times to assist any Army patrol that might run into trouble. These FACs were also conducting reconnaissance in the AR looking for targets as well as being available if called upon.

Additionally, some of the Ho Chi Minh Trail infiltration routes from North Vietnam through Laos had entry points into South Vietnam in our AR. We would interdict these entry points each day just before sunset (See Photos 13 and 14). Our goal was to close down the trails into South Vietnam or at least slow down the North from bringing supplies into the South each night to resupply the Vietcong (VC or "Charlie"). This was accomplished either by air strikes or the use of the Army's 175 artillery tubes. Then each morning we would do a "first look" at each of these interdiction points to see just how much truck traffic had occurred during the night.

On each of these "first look" sorties there was always evidence of truck traffic passing over these trails bringing in supplies from the North. The North Vietnamese worked very hard keeping "Charlie" supplied in the South and building up stockpiles for any future offensive actions planned against the South.

Our daily FAC operations out of Phu Bai consisted of supporting the Army with close air support missions, directing scheduled air strikes on fixed targets, interdiction missions and reconnaissance missions.

After my arrival at Phu Bai and the two day AR orientation sorties, I was cleared to fly missions in support of the Army. A very interesting thing happened on the 2nd or 3rd mission that I flew (See Photo 15). I was out in the western part of the AR, in the A Shau Valley, when a voice came up on my ultrahigh frequency (UHF) radio. The voice spoke in broken English and welcomed me to Vietnam. My call sign at that time was "Bilk 28." This voice said, "Welcome to Vietnam Bilk 28, we are going to shoot you down." This was a little unsettling to say the least. Once again, I thought to myself, this is going to be a very long year.

When I got back to Phu Bai, I asked Major Robbins (Robby) if this was a normal occurrence. He responded by saying, "Oh yes, Charlie welcomes all of us FACs to Vietnam." It also told me that our enemy had a very significant intelligence network in place. They knew I was new to Vietnam, and they knew my call sign. They also knew our radio frequencies. One has to wonder how they got this type of information. One also had to be impressed that they got the information so fast.

I now understood why Robby had stressed that I should always use a different departure route after takeoff when departing Phu Bai. It was obvious that "Charlie" was monitoring our takeoffs and landings as well as our departure routes. Being random in departure was the best way to avoid being predictable, which could lead to being a good target on takeoff. I never forgot this lesson, especially after my arrival on the Continental flight and the 50-caliber round through the horizontal stabilizer on landing. This lesson served me well in Vietnam.

The O-2A has two engines in a push/pull configuration. The pull engine is on the front of the aircraft while the push or rear engine is located behind the cockpit between the two booms of the tail section. With both engines operating, one can achieve altitudes up to approximately 8,000 to 10,000 feet for sustained level flight. If you lost the rear engine, you could only maintain a maximum altitude of approximately 1100 feet for sustained level flight. During my tour in Vietnam, I lost the rear engine on three different missions. I only lost engines twice during the rest of my 25-year Air Force career.

The first time I lost the rear engine in Vietnam occurred on a mission just after "Charlie" had welcomed me to Vietnam. I was out in the A Shau Valley on a normal reconnaissance mission when I lost oil pressure on the rear engine. The engines on the O-2A automatically shut down if oil pressure is lost. The A Shau Valley is surrounded by mountains that range from 3000 feet to 5000 feet in altitude. There is only one pass out of the A Shau Valley where the ground level is below 1000 feet.

When I lost the engine, I was near the southern end of the Valley; and the pass is near the northern end. As I began the mandatory descent to a level that I could maintain level flight, I headed back to the only area that would allow me to exit the Valley. As I proceeded through the pass to the east out of the valley, I noticed 50-caliber traces outside the cockpit coming from above me from the top of a mountain, and I wondered if "Charlie" was going to make good on his welcoming greetings of shooting me down.

Tracer rounds are used every fifth round. I saw a number of tracers coming from above me and passing below. Thankfully, I made it through the pass and back to Phu Bai without being hit. That was my first mission on which I took ground fire but definitely not the last. Again, I thought to myself this really is going to be a very long year.

It was on August 8, 1971, 16 days after my first flight in Vietnam, when I was awakened at 2300 hours (11:00 P.M.) by Maj. Robbins (Robby). Robby told me that we had a very critical "troops in contact (TIC)" situation at Fire Base (FB) Sarge which was located in the most western section of Quang Tri Province, in the 1ˢᵗ Brigade's AR. I reminded Robby that I had not had my night checkout mission and that I had not flown in Quang Tri Province area since that AR belonged to the FACs at Camp Evans. He responded by saying "Yes, I know, but you are the only one who has proper crew rest among our pilots and the ones at Camp Evans." He then said, "The situation is critical and you have to go." I guess he figured that since I was a mid-range Captain and had previously been an Instructor Pilot (IP), I could handle it. I was also the next ranking officer at this location. I immediately got up and rapidly got dressed for flight. Once again, the thought crossed my mind that this is going to be a very long year.

While I was getting dressed, Robby got our maintenance personnel to get an aircraft ready for flight. Robby met me at the aircraft and said that I would not have any problems finding FB Sarge. He said, "Just fly up the coast to Quang Tri City, look west, and where you see all of the action going on—that's FB Sarge." Robby was correct. I had no problem finding FB Sarge. It looked like a 4th of July celebration with fireworks. I arrived on station 45 minutes after being awakened by Robby.

When I arrived on station, the Army was providing tactical fire support with their 175 artillery batteries. I contacted the ground commander at FB Sarge, and he provided me with all of the information I needed to take over control of the close air support. I then contacted the 175-artillery battery commander and began directing his fire support on the areas where the North Vietnamese Army (NVA) and/or the Vietcong (VC) were attacking the FB. FB Sarge was being attacked on three sides. I immediately knew that I would need more fire power, so I contacted our Direct Air Support Center (DASC) at Da Nang and requested any air support that was available for a "troops in contact" (TIC) situation. The Air Force normally had air strikes in progress each night over the Ho Chi Minh Trails in Laos, which were not far from our location. I also knew that we had F-4 fighters on alert at Da Nang.

It was not long before an AC-130 gunship checked in with me from off the trails. I held up the 175s and began directing the AC-130 on the areas of most concern. When the gunship was "Winchester" (out of ammo) he departed for home. I then returned to the 175s to keep ordinance falling on the "bad guys" while I waited for additional air support.

It was not long before the first set of F-4s from the Gunfighter Squadron at Da Nang checked in with me. I put these F-4s in close to the FB as the "bad guys" were getting very close to the FB's perimeter. During the next three and a half hours, I was provided with three additional sets of fighters and another gunship. So, for four and a half hours, I directed four sets of fighters, two gunships, and 175 artillery tubes in support of FB Sarge, which did result in finally breaking the "troops in contact" situation. After four and a half hours "Charlie" decided he had had enough and retreated back into Laos. That was very good for FB Sarge and good for me, since I was getting low on fuel and I was almost out of marking rockets. We had turned back "Charlie" and kept him from overrunning FB Sarge that night. I returned to Phu Bai just as the sun was beginning to rise.

Robby met me at the aircraft with a big smile on his face. He had been monitoring the action from our Tactical Operations Center (TOC) and was fully aware of what had happened. As I got out of the aircraft, I asked Robby if I should call Da Nang and tell them that I had my night checkout. He just smiled and said, "By all means."

The next day, FB Sarge called our TOC and provided a bomb damage assessment (BDA) for the previous night's operation. They reported 258 "bad guys" killed in action (KIA) and also said that "Charlie" had dragged off many more.

I was awarded the Distinguished Flying Cross (DFC) medal for that mission. I was just happy that I was able to support our troops on the ground that night. I also learned some very valuable lessons. I learned how vicious war actually is and what intense automatic weapons ground fire looks like. Being a fighter pilot and going fast, dropping bombs and firing guns is fun, but there is no greater feeling than knowing that you were responsible for providing the means that actually saved the lives of our troops on the ground. I knew my actions had saved many lives that night at FB Sarge by me just doing my job.

The next two months were not as exciting as my night self-checkout. The daily missions became routine. I flew combat patrol, interdiction and reconnaissance missions. In addition, the Army would conduct search and destroy missions. We would also participate in these types of missions which consisted of the

Army identifying a relatively large area as a "free fire zone". These areas were identified as suspected enemy held areas, and all of the required clearances would be obtained for a specific period of time prior to takeoff. These missions consisted of what the Army called "Pink Teams". These teams consisted of a "Little Bird" (See Photo 16), two Cobra gunships and a "Command and Control" (CC) Huey helicopter. A FAC would also accompany these teams on the search and destroy missions into the designated "free fire zone."

These missions consisted of "Little Bird" flying on the deck just above the treetops, trolling for enemy gunfire. The crew of this helicopter consisted of a pilot and a gunner sitting at the door with his feet on one of the landing rails. In one hand, the gunner had an M60 machine gun; and in the other hand, he held a smoke grenade. Flying just above "Little Bird" were two Cobra gunships that were flying in circles on either side of "Little Bird's" flight path. These circles were coordinated to ensure that at least one of the Cobra gunships was in a position to respond if "Little Bird" took fire. Above these three helicopters, the Huey Command and Control (CC) helicopter was positioned. His function was to be in charge of the mission. Above these four helicopters, in this moving stack of aircraft, was the FAC. When "Little Bird" took fire, the gunner would immediately throw the smoke grenade and start firing the M60 machine gun. When the smoke grenade ignited, the gunships would immediately roll in on the target as "Little Bird" exited the area. If the target turned out to be too much for the gunships to handle, the target was turned over to the FAC for engagement. The FAC would order an air strike through the TASC who in turn would contact the DASC at Da Nang for support. The target would then be serviced with fighter aircraft. This was a very effective way to neutralize targets and prosecute the war.

I did have an interesting mission on one of the daily "first look" sorties to check out the interdiction points. I discovered a truck that had slid off of the trail and was stuck. What probably happened was that after the Vietcong (VC or "Charlie) had cleared the trail, this truck was probably trying to beat sunrise. It most likely took

19

"Charlie" longer to open the trail after our closure the night before, and this truck was going too fast as it was trying to get through this interdiction point before sunrise. Once past these points, the trucks would disappear into the triple canopy jungle.

On this day, one of the truck's back wheels slipped off of the trail; and the truck became stuck in the open area of the interdiction point. I spotted the truck, and the driver was trying to conceal the vehicle by shoveling dirt on top of it. He was never going to be able to cover up the truck, but at least he was trying.

I contacted our TASC who contacted the DASC at Da Nang and requested air support. About 20 minutes later two F-4s from the Gunfighter Squadron at Da Nang checked in with me. I briefed them on the target and told them that I would mark it. Marking the target probably was not necessary since the truck was very obvious against the bare soil from the repeated closures each night. It was, however, always fun to shoot rockets, and the smoke would assist the fighters by giving them the existing wind conditions.

When the smoke rocket detonated, the truck driver disappeared into the jungle. I decided to have the fighters drop one bomb at a time since I had not observed any ground fire coming from the target area. Lead dropped his first bomb close to the truck but did not destroy it. Two rolled in; and when he released his bomb, it landed on the covered bed of the truck and exploded. The truck tumbled down the side of the mountain and was no longer a concern. The fighters wanted to know where I wanted the remainder of their bomb load placed.

Since we felt that the North Vietnamese Army (NVA) or the Vietcong (VC or "Charlie) used heavy equipment, like bulldozers or front end loaders to clear the trails each night, I decided to try and get lucky. We felt that this heavy equipment was probably located near each of the interdiction points and concealed by the triple canopy jungle. I directed the fighters to drop their remaining bombs in the jungle area near the trail and the interdiction point. Who knows, we may have gotten lucky and actually found the needle in the haystack.

While at Phu Bai, I also learned another valuable lesson. I learned to never trust any of the Vietnamese working on base. Supposedly, these workers had all been vetted, but how can one really be sure where these workers' allegiance actually lies.

This lesson was driven home one night around 0200 hours (2:00 A.M.) when the base came under a rocket attack. In addition, "Charlie" was probing our perimeter razor wire fence. After a short time of small arms fire near the wire, the probe ceased, as did the rockets. The next morning two of the individuals who had been doing the probing had been killed and were lying in the wire. One of these individuals happened to be one of our Vietnamese barbers on base. He cut our hair during the day; and then at night he became "Charlie" trying to kill us. You never really knew who the enemy was.

I also saw the good and the bad of the Army while at Phu Bai. The good was displayed by the actions of many of the helicopter pilots. It was on one of the search-and-destroy missions that I witnessed the loyalty and the strong bond that is created between fellow soldiers during war.

On this particular mission, "Little Bird" was shot down and crashed on top of the triple canopy jungle. When it crashed, the small chopper seemed to roll like an egg on top of the triple canopy. Since "Little Birds" are shaped like eggs, this crash pattern was not unexpected. What was unexpected was when the Huey CC bird descended to the crash site to check on the crew, at about 15-20 feet above the crash site while in a hover, one of the crew members of the CC ship jumped out of the helicopter without a parachute. He landed on top of the triple canopy jungle and made his way over to "Little Bird" where he pulled both crew members out. I watched in amazement at this unselfish and brave act. Unfortunately, both crew members had been killed in the crash. When asked later why he had jumped out of the CC Bird, this individual merely responded saying, "I had to. The pilot of 'Little Bird' was my buddy." That is what dedication, bravery, and loyalty is all about. The bond developed between military members in wartime is extremely strong. One has to respect it.

Another good example of the dedication, bravery and loyalty of these young 18-and 19-year-old helicopter pilots was displayed when an Army patrol team out in the A Shau Valley west of Phu Bai became pinned down by enemy fire. The Army TOC launched a number of Cobra gunships to provide close air support to these troops.

The report of "troops-in-contact" (TIC) came in just after sunrise. A number of gunships were launched to address the situation. At approximately 0700 hours (7:00 A.M.), one of the gunships was shot down. A rescue Huey Medevac helicopter ("Dust Off") was able to retrieve the pilot and gunner and bring them back to Phu Bai. The pilot immediately went to the Flight Surgeon and requested to be cleared for flight. Fortunately, he was not injured and the Flight Surgeon did clear him for flight. He immediately came back to the TOC and requested to go back out to the site since the situation had not been resolved. The Army Brigade Commander authorized his return to the battle in another gunship.

At approximately 1200 hours (12:00 P.M.), he was shot down again; and again "Dust Off" was able to rescue him unhurt. He again went to the Flight Surgeon and requested to be cleared for flight. The Flight Surgeon once again cleared him, and he returned to the TOC. This time the Brigade Commander sent him out in a Huey Command and Control (CC) Bird to reenter the battle area.

As bad luck would have it, at approximately 1600 hours (4:00 P.M.) his CC Bird was shot down. He and the crew were again rescued unharmed and returned to Phu Bai. This time after the Flight Surgeon cleared him for flight, the Brigade Commander refused to let him go back out to the site. The Brigade Commander said, "Three choppers in one day are enough. I can't afford you flying anymore today." The young Warrant Officer did not like this answer, but I think he understood. I do not know if this was a record or not being shot down three times in one day, but it did exhibit the type of courage and bravery that many of these young Army helicopter pilots had.

In addition, many of these young Warrant Officers had a goal of getting 1000 hours of flight time during their year in Vietnam. If they did, they could return to the States and get a good paying job as a helicopter pilot for a radio station news department or a job flying helicopters transporting patients. They would do anything to get flight time.

I observed an example of one of these young pilots trying to get as much flying time as possible. I happened to be getting my annual flight physical when a young Army Warrant Officer pilot poked his head into the Army's Flight Surgeon's office and asked if he could be released from DNIF status (Duty Not Involving Flight). This is a status all of us pilots hate. The Flight Surgeon looked at him and asked, "How high can you raise your arm?" His right arm was in a sling. The younger officer slipped off the sling and tried to raise his arm. He could only raise it up to about shoulder level. The Flight Surgeon looked at him and said, "Not today son." The young pilot put the sling back on and went on his way. As he departed he said he would be back tomorrow.

The Flight Surgeon then turned to me and said that he had taken 12 or 13 pieces of helicopter canopy Plexiglas out of the young pilot's arm the day before when he had been shot down. These young chopper pilots had no fear, and they felt that they were bulletproof. I guess that is why the Army liked these young pilots. These young pilots had very high morale while the Army ground troops that manned the FBs for long periods of time had very low morale.

In addition to the good I also witnessed the bad. During this time period, outside of the helicopter pilots, the Army in general was experiencing very low morale as well as a significant drug problem among its enlisted members.

As I previously mentioned, our Tactical Operations Center (TOC) was co-located with the Army's TOC for its helicopter assets. The Army manned their half of the TOC 24 hours a day. We manned our half only during the time we had flight operations going on. Each night, however, I had to go over to the TOC to get the next day's Fragmentation Order (FRAG) which specified the next day's scheduled air strikes in our AR. Since I was the second ranking officer, Major Robbins (Robby) had put me in charge of scheduling. It was my job to schedule our daily flight operations.

The TOC was located in a secured bunker, which used a daily password for entry for security purposes. In order to enter the TOC, the daily password had to be used. Initially, when I took over flight scheduling and I would go to the TOC at 2200 hours (10:00 P.M.) each night to get the FRAG, I did not notice any potential problems. Then as time passed, I started noticing small empty plastic vials lying next to the PSP walkway leading to the TOC. At first it was just one or two of these empty vials. Not knowing what these vials were, I asked. I was told that these vials, which were in two sizes, were nickel and dime bags for drugs. This then became a big concern for me. The Army leaders seemed to be aware of the problem but seemed to turn their heads as if not wanting to know. I, of course, had a great concern since I had to enter the TOC each night at 2200 hours (10:00 P.M.). To make matters worse, the number of empty vials was increasing over time.

To address my concern and even though the Army personnel manning the TOC knew that I would be coming, I would make as much noise as I could while approaching the TOC. I would whistle or walk heavily on the PSP just to let them know that I was approaching. This gave them enough time to put away anything that they did not want me to see and anything that I certainly did not want to see. I also hoped that they would remember the password for the day. The number of empty vials never decreased during the time I was at Phu Bai.

I was also there when an incident of fragging occurred. Fragging is when someone would open one of the doors on either end of a living quarter's hooch and toss a live hand grenade inside. It was usually an action being accomplished by one Army individual against another. It usually involved an enlisted member fragging an officer for whatever reason. I am sure that drugs were a contributing factor to this type of activity. Fortunately, the individual that was fragged while I was there was not killed; however, no one was ever identified that I know of as the fragger. Morale was not high in the Army during the latter part of 1971.

Another one of a kind mission that I flew in support of the Army involved the creation of an instantaneous five-ship landing zone (LZ). The Army had decided to enter an area that was known as a VC stronghold area and the Army had not paid too much attention to in the past.

The Air Force had a few 15,000-lb bombs available, that when dropped out of the back of a C-130 aircraft, would create an instantaneous five-ship helicopter LZ. A six-foot fuse extender was attached to the nose of the bomb, and then it would be dropped out of a low-flying C-130 aircraft with the use of parachutes. Three large parachutes would slow the descent of the bomb and orientate the bomb nose down as the C-130 exited the area. I had never seen one of those bombs dropped before.

I was briefed before takeoff as to where the Army wanted this LZ to be created. After airborne, I made contact with the C-130 aircraft and directed him into the target area. I marked the location with a smoke rocket and confirmed with the C-130 that they had the location. I then climbed up to an altitude of approximately 4500 feet and moved away about four to five miles from the target area. Based on my experience with 2000-lb bombs, I felt that this would be sufficient. I could not have been more wrong.

I watched as the C-130 started its run toward the target area. He was flying at an altitude of about 2000 feet and was well below me. I watched as the rear cargo door opened and the small drag chutes deployed as the C-130 approached the target area. These drag chutes pulled the 15,000-lb bomb out of the aircraft, and then the three large main chutes deployed to slow the rate of descent of the bomb. The bomb descended nose down while being held up by the three large chutes. This nose-down descent allows the six foot fuse extender and the fuse to contact the ground first when the bomb is still six feet above the ground. The blast from this bomb creates an instantaneous five-ship LZ without creating much of a hole.

As I watched the bomb slowly descend and the C-130 move out of the way of the bomb blast, I felt comfortable with my position. As the bomb detonated six feet above the ground, I could see the shock wave and the blast effect, which looked like the pictures of nuclear weapons going off when these weapons were being tested. When the shock wave from the blast reached my altitude and distance away, my aircraft and I were turned every way but loose. I recovered the aircraft inverted in a 25-degree nose low attitude. I was very lucky that no structural damage had occurred to the aircraft. I learned a very important lesson about 15,000-lb bombs— they are big! The Army was very happy with their new LZ.

I had been at Phu Bai for about two months when the 504th TASG received a message from the O-2A Training Wing at Hurlburt Field in Florida where all of the O-2A pre-Vietnam flight training was accomplished. This message was an inquiry about what the 504th TASG thought about the quality and effectiveness of the training being provided to FACs on their way to Vietnam. The 504th TASG asked all three of its squadrons to provide input to them to assist them in their response.

Each TASS was to ask a newly arrived FAC for his opinion on the quality of training he had received and how well it prepared him to accomplish the mission in Vietnam. Robby asked me to respond since he knew that I had previously been an Instructor Pilot (IP) for Undergraduate Pilot Training (UPT) for the Air Force. He felt that since I had just recently completed O-2A training, and with my IP experience, I would be the best person to evaluate the quality of training being provided in Florida and its effectiveness in Vietnam. This was especially true since most of our FACs were young Lieutenants, and did not have much experience to begin with.

It probably was unfortunate for the Hurlburt Field Training Wing that I was selected to provide input. I had not been impressed with the O-2A training that I had received there. I felt that a number of things were lacking in my preparation for combat operations in Vietnam. Having been an IP for four years before Vietnam, I had the experience to overcome these shortcomings. I did however, have a major concern for the recently UPT graduates that received FAC assignments upon graduation. I did not feel that these young pilots had the experience to overcome some of the shortcomings that I had experienced during O-2A training.

As I have mentioned, I was the second ranking officer, as a mid-range Captain, at Phu Bai, and we had 10 pilots assigned. Major Robbins, the ALO, was the senior officer at Phu Bai. At Camp Evans, a similar situation existed. In fact, my Flight Commander from Laughlin AFB, a Major, was the ALO at Camp Evans. He had left Laughlin AFB two months before me. Most of his 10 pilots were also recent UPT graduates. For the most part, the FACs in Vietnam during 1971 and 1972 were very young and inexperienced pilots.

One of my major concerns about the O-2A training that I received was due to a condition that was actually beyond the control of the Training Wing at Hurlburt Field. This had to do with the physical location of the O-2A training being conducted in the Florida panhandle. This physical location made it almost impossible to teach topographic map reading skills for use in Vietnam. Topographic map readings were probably the most essential training requirement needed to be an effective FAC in MRI and MRII in Vietnam.

Hurlburt Field is located on the Florida panhandle next to the Gulf of Mexico. There essentially is no topographic relief present there for hundreds of miles. In addition, this area is heavily populated with roads, highways, towns and many more man-made features along this coastal area. This training area was totally different from the topography in Vietnam, especially in MRI and MRII. These two regions represented two thirds of the country where Air Force FACs were operating.

In these two regions, the terrain is primarily mountainous and/ or covered with triple canopy jungle (See Photos 17, 18, 19, 20, and 21). There are very few roads or man-made features in the mostly uninhabited area where "Charlie" was located. In MRII, this region was called the Central Highlands, and it contained the highest mountain peak in Vietnam at a height of 8,524 feet. It was impossible to teach map reading skills to young pilots in Florida for use in Vietnam in MRI and MRII.

I also knew that map reading was not emphasized in UPT since this skill is not really required for most of the missions in the Air

Force. Topographic map reading skills, however, was mandatory for FACs in MRI and MRII. The FACs in MRIII and Cambodia did not have as much of a disadvantage since this region was mostly flat and more densely populated. This area had many roads and other man-made features for use. It was somewhat comparable to the terrain around Hurlburt Field. This concern was more of a problem for the Air Force rather than for the Training Wing at Hurlburt Field. The Wing did not have any choice as to where the O-2A training was to be conducted.

There was a good aspect for conducting the O-2A training at Hurlburt Field. This was due to the fact that the Air Force had a number of fighter aircraft bases and ranges nearby, which allowed for excellent training of controlling fighter aircraft as a FAC. There were fighters based at McDill AFB in Tampa as well as Eglin AFB at Fort Walton Beach in Florida.

My second major concern was more of a personal problem for me. The IP that I was assigned to for O-2A training as an old Major who had been passed over for Lieutenant Colonel (LTC) a number of times and was just putting in his time prior to retirement.

Many times my training missions with this IP consisted of takeoff and then flying directly to the coast to see where the fish were running so that he could go fishing after he finished for the day. We would make two passes up and down the coast between Fort Walton Beach and Pensacola looking for fish. He would then have me take him back to Hurlburt Field where he would get out, and I would then go solo to complete the mission.

Having been an IP myself, this did not impress me. I was doing a lot of self-instructing during my O-2A training in Florida. This was O.K. for me because of my experience; however, it would not be satisfactory for a new pilot right out of UPT. I knew that I was not his first student, and I knew I would not be his last. This type of training, in my opinion, would be totally inadequate to prepare a young pilot to be a FAC in Vietnam.

Robby forwarded my response to the 20th TASS who forwarded it on to the 504th Group for their use in preparing their response to

Hurlburt Field. Two things happened after I provided my input to the 504th Group. The first thing was that I was asked by the 504th Group if I would be willing to go to Nha Trang AB and set up an in-country training program for all new FACs arriving in Vietnam. They felt that with my background as an UPT Instructor, I was the perfect one to set up such a program. They seemed to agree with me that the O-2A training at Hurlburt Field was inadequate, and they decided to do something about it. I agreed, and after two months at Phu Bai with the 101st Airborne, I was on my way to Nha Trang to set up a new in-country FAC training program.

The second thing that happened was that after I got to Nha Trang, a full Colonel from Hurlburt Field arrived to discuss my concerns about their training program. So there I was, a mid-range Captain telling this full Colonel how he could improve his training program. Usually, it is the full Colonel telling the Captain how the cabbage is eaten.

Air Force must have felt that my concerns were serious enough to send a full Colonel, in person, to Vietnam to listen to my concerns. I was totally surprised, but impressed, that those in a position to do something about my concerns that could mean life or death for a young pilot, were willing to listen and make changes. By the time I left Vietnam, I did see a lot of improvement in the quality of the new arriving FACs. The stateside training program had been improved.

CHAPTER FIVE

Nha Trang Operations

IT WAS OCTOBER 1971, and I was now on my way to Nha Trang Air Base. Nha Trang is located approximately 300 miles south of Phu Bai in MRII on the coast (See Figure 2). It is situated in a valley surrounded by mountains. The 21st TASS had an FOL at this location, which provided tactical air support for MRII. I was now assigned to the 504th TASG since I was going to be providing the training for all new arriving FACs. These new FACs would be assigned to one of the three TASSs after completing this new in-country training program.

I reported to a Lieutenant Colonel (LTC) MacPhearson at the 504th Group. He also had been an UPT Instructor Pilot (IP), so we had a lot in common. Since the idea of this in-country training at a level that I thought was necessary to overcome the basic O-2A training deficiencies, LTC MacPhearson essentially gave me the lead. He said he would supply me with whatever resources I needed to get the job done. This was great knowing that I had the full support of the 504th Group, which would be needed in order to develop an effective in-country training program. The 504th Group also moved some additional aircraft and maintenance personnel to Nha Trang to support this new mission. These assets were assigned to the 21st TASS since the FOL was part of the 21st TASS's area of operations. The 504th Group remained stationed at Cam Ranh Bay.

I developed a syllabus for the O-2A in-country training, which consisted of 10 sorties with emphasis on developing map reading skills. MRII was very mountainous and covered with triple canopy jungle. LTC MacPhearson approved my syllabus and said that he would help me fly the training sorties. He even moved from Cam Ranh Bay to Nha Trang to assist me.

As I moved south, my living conditions vastly improved. The large bases like Cam Ranh Bay, Nha Trang and Phan Rang were very much like stateside Air Force Bases.

During this time period, President Nixon's Vietnamization Program had been progressing, and many of our U.S. forces had stood down and were returning to the States. With their departure, the 504th Group had access to better flight line facilities and aircraft parking areas for our aircraft. These facilities had previously been utilized by the departed fighter squadrons. For example, at Phu Bai our aircraft were parked in sand-filled bunkers located on PSP decking (See Photo 22). At Nha Trang, our aircraft were parked in the same type of sand-filled bunkers but were located on a concrete tarmac (See Photo 23). Our flight line office facilities were also much nicer and fully air conditioned with running water, restrooms and continuous electrical power.

My living quarters at Nha Trang were also much better than the ones at Phu Bai. We had the quarters that the fighter squadrons had lived in. These units had electricity all of the time, hot and cold running water and indoor latrines. Nha Trang was much like a stateside Air Force Base. It even had an Officers' Club and a Base Exchange (BX) that was always well stocked. The Air Force just knows how to live better than the Army.

At Nha Trang and its Officers' Club, we could get a decent meal. At Phu Bai, we FACs did get to eat meals with the Brigade Commander, so we did not have to go to the Army's Mess Hall. This was okay but while I was at Phu Bai, I would use any excuse to divert to Da Nang if it was around mealtime. At Da Nang we could go to the Navy Mess, which had excellent food. The Navy eats well.

It was also during this time that the war had entered into a period of much lower intensity as far as the U.S. forces were concerned as Vietnamization took effect. FACs were still providing tactical air support in all three MRs as well as half of Cambodia, but at a much lower level of intensity.

It took approximately two weeks to develop the training syllabus and get it approved. I also set up a training flight operations area on the flight line in one of the departed fighter squadron buildings. I was informed that I would receive my first group of new FACs the next week. I now had a week to set up housing and line up transportation for use during the training.

Since the Air Base had most of the support facilities as stateside bases, I contacted the base transportation unit to arrange for transportation to pick up the new FACs when they arrived. This transportation unit was small and they indicated that they may not always have a driver available on short notice. I asked what would be the best solution for me to cover my transportation needs. They suggested that I get certified to drive one of their buses and then I could check out a bus whenever I needed it for as long as I needed it. This sounded like a good solution.

The bus certification consisted of me taking a checkout ride with one of their drivers. The driver that checked me out was an Airman First Class. The checkout consisted of me driving the bus around Nha Trang Air Base while he observed. He got a kick out of riding in the bus with a Captain driving it. He would have me honk at people he knew while he waived as we drove around the base. He then stamped my military driver's license with "48 Passenger Bus." This turned out to be a very good solution, and transportation was never a problem.

My first class consisted of five newly arrived FACs. LTC MacPhearson and I essentially flew two sorties a day as we provided the newly developed in-country training for these new FACs. After completion of this in-country training, these new FACs were assigned to one of the three TASSs of the 504th Group.

I was at Nha Trang for approximately two months providing this new in-country training for new FACs when the 504th Group decided to move this training activity to Phan Rang Air Base. As the U.S. forces were continually being drawn down, the F-100 Fighter Wing stationed at Phan Rang was sent home. Phan Rang AB was even larger than Nha Trang, and now the vacant fighter facilities offered an even better environment for training.

Phan Rang was located approximately 70 miles south of Nha Trang. I was now on my way to Phan Rang. At least I was moving farther south, and with each move my living conditions improved.

CHAPTER SIX

Phan Rang Operations

I ARRIVED AT PHAN Rang around the 1st of December 1971. The living conditions once again greatly improved. Phan Rang was even more like a stateside AFB than Nha Trang (See Photo 24). We moved into the housing units and flight line facilities that previously housed the F-100 fighter squadrons (See Photos 25 and 26). The living quarters were like an 18-unit dormitory type facility with a central latrine area (See Photos 27, 28, 29, 30 and 31). We also had the previous F-100 flightline facilities from which to conduct our flying activities. These were very much like stateside facilities. In addition, the runway was long and could handle all types of aircraft. We had plenty of room for our aircraft and maintenance facilities. Phan Rang was also located on the coast, and there were fewer mountains immediately around the base.

There were only two negatives about Phan Rang as far as I was concerned. The first was I had to walk by the Army's mortuary each morning on my way to the flight line. This mortuary was the U.S. casualty unit that prepared our fallen soldiers for their final trip home. Walking by this location each day was sobering because I could see the stacks of wooden coffins ready for use. There were always a large number of these coffins, and it was very obvious that these coffins were being utilized on a daily basis. There was always a different number of stacks and each stack always had a different number of coffins just waiting to be used. It was obvious that use and resupply was occurring daily. Additionally, the smell of formaldehyde was very strong. It seemed to take an hour or two each day for me to get the smell out of my mind. This was not a good way to start each day as I headed to the flight line to fly.

The second negative about Phan Rang was when I discovered one of the roads on base had been named after one of the F-100 pilots that had been killed. This pilot was one of those names I had seen on the weekly lists as Killed in Action (KIA) during my fighter qualification training at Cannon AFB prior to my departure from the States. This pilot was also my neighbor and one of my fellow IPs at Laughlin AFB. His name was Mike McGovern, and we were in the same flight at Laughlin AFB.

This road was on a small hill near the center of Phan Rang and was used to get to the top of the hill where the Base's radio towers were located (See Photos 32 and 33). Every time I looked at this hill, it brought back the memory of my friend, neighbor, and fellow flight member.

Just before the end of December of 1971, President Nixon's Vietnamization Program was progressing rapidly. We were notified that the 504th Group was going to be deactivated along with the 19th TASS and sent home. Anyone with less than four months remaining on their tour was going home early. Since I still had six months to go, I was not one of those going home. LTC MacPhearson did have less than four months and he was going home. I was going to be assigned to the 21st TASS at Tan Son Nhut Air Base at Saigon in MRIII. I was again moving south. This ended my in-country training program. I did learn later that my training program had been affectionately named FAC-U by recent graduates. It was known as the Forward Air Controllers "University" at Phan Rang.

CHAPTER SEVEN

Tan Son Nhut Operations

I ARRIVED AT TAN Son Nhut Air Base in January 1972. The Vietnamization Program was accelerating. U.S. strength levels in January 1972 were down to approximately 140,000 troops. This was down from a peak level of over 560,000 troops. It was also projected to be down to below 70,000 troops by April 1972. The war was being turned over to the South Vietnam's Army of the Republic of Vietnam (ARVN), the Vietnamese Marine Corps (VNMC) and the Vietnamese Air Force (VNAF). We still had Army and Marine advisors with Vietnamese units, but many of our combat ground forces were standing down.

In MRI the 101st Airborne Division had stood down and was sent home. MRI had been turned over to Vietnamese forces. There were still a few of our ally forces present like the South Koreans and our Special Forces (SF) operating in MRII in the Central Highlands north and west of Nha Trang. The South Korean forces were excellent fighters and very capable of defending their area of responsibility (AR). Our Special Forces (SF) continued reconnaissance missions in their area of responsibility (AR). They were operating out of Ban Me Thuot in MRII.

With the departure of the 504th Group and the deactivation of the 19th TASS, the U.S. had only two Tactical Air Support Squadrons

(TASSs) remaining in-country. The 20th TASS remained at Da Nang and still provided the tactical air support required for MRI. The 19th TASS was essentially combined with the 21st TASS which was now located at Tan Son Nhut. The 19th TASS had previously been responsible for tactical air support in MRIII and Cambodia. By combining the 19th TASS and the 21st TASS assets into one large squadron, the 21st TASS picked up the responsibility for tactical air support operations in MRII, MRIII and approximately one half of Cambodia (See Figure 2). This became a very large squadron with six FOLs located in MRII and MRIII.

Air Force organizational structure does not allow for standalone squadrons. With the 504th Group's departure, the 20th and 21st TASSs needed to be assigned to Air Force Wings for administrative support. The 20th TASS was assigned to the Wing at Da Nang while the 21st TASS was assigned to the Wing at Tan Son Nhut. The Wing that the 20th TASS was assigned to at Da Nang was a Fighter Wing and therefore had similar missions. The Wing to which the 21st TASS was assigned to at Tan Son Nhut was an Airlift Wing. The 21st TASS was a combat operational squadron assigned to an Airlift Wing that had totally different missions. This turned out to be a very serious potential problem for me.

Upon my arrival at Tan Son Nhut and the 21st TASS, I was not sure how I would be utilized. LTC MacPhearson, who I had worked for and with at Nha Trang and Phan Rang had gone home with the 504th Group. I essentially was an unknown quantity to the 21st TASS; however I still had six months to go on my tour. I reported into the 21st TASS and waited to see what I was going to do the remaining six months of my tour. It was not long before I found out just what was in store for me at the 21st TASS.

As I mentioned earlier, the ranks of the FAC corps were very young and inexperienced. After I reported in, I found out that the 21st TASS Squadron Commander was a LTC from the Strategic Air Command (SAC). This LTC had been a bomber pilot his entire career, and he was not happy being a FAC. The Maintenance Officer was a Major and was going to retire upon rotation back to the States. He also was not very enthusiastic about being a FAC. I then found out that I was the highest ranking Captain, and, as such, I was going to be the Operations Officer for the 21ˢᵗ TASS. This was quite a surprise for me as this is usually a Lieutenant Colonel's position.

I was now the Operations Officer of the largest tactical air support squadron in Vietnam. We had combat tactical air support responsibility for MRII, MRIII and one half of Cambodia. We had approximately 80 pilots, 40 aircraft and 235 Maintenance personnel assigned to provide tactical air support for a very large portion of Vietnam and Cambodia. This support was being provided out of seven different operating locations in Vietnam. We had this responsibility while being assigned to an Airlift Wing, which operated under a totally different set of operating procedures. We essentially did not speak the same language when it came to flight operations and missions. The situation that I found myself in was that I had a Squadron Commander who had flown B-52s all of his career and who was not even sure how to spell FAC, let alone be one, assigned to an Airlift Wing whose flight operations and missions were totally different from combat operations. I soon found out that I was in a potentially impossible situation.

My Squadron Commander told me that he would take care of all of the paperwork, and I was to take care all of the flying operations including briefings to 7ᵗʰ Air Force. In Vietnam, 7ᵗʰ Air Force had responsibility of all flying operations and they were the ones that issued the daily FRAG Order for combat operations.

I also did not see much of the Major Maintenance Officer as he kept a very, and I mean a very low profile. He was only putting in his time until he went home and retired. He never flew missions, but he did do a very good job of keeping our aircraft flying. The Squadron Commander also did not fly any combat missions.

As Operations Officer, I now had the responsibility to brief 7ᵗʰ Air Force on the progress of the war from our perspective for the areas of our responsibility. 7ᵗʰ Air Force had direct control over all combat flight operations in this theater of operations including tactical air support. These people were in charge of the Air War in Vietnam.

Since we were assigned to the Airlift Wing, as TASS Operation Officer, I reported to the Airlift Wing Operations Officer who was a full Colonel. I was a mid-range Captain who was filling a position normally held by a LTC and was reporting to a full Colonel.

Our initial meeting did not go very well for me. After I introduced myself, we began to discuss FAC operations. It soon became apparent that this Colonel had no concept of what FACs actually did since he had been an airlift pilot his entire career.

He first asked me about how we scheduled our flights and what type of operational control we had over our airborne aircraft. I explained to him that besides our main operating location here at Tan Son Nhut, we also had six FOLs located in MRII and MRIII that we operated out of daily. I explained to him that each location essentially operated autonomously and prepared and flew their own schedules in support of their areas of responsibility. I further explained that each location had a Tactical Operations Center (TOC) that controlled their own flight operations. This seemed to raise a question in his mind.

He then asked me how many aircraft did I have airborne at that very minute. I replied that I did not know specifically how many right then, but there were probably five or six airborne in MRII and MRIII and two airborne in Cambodia. He then asked me how we kept track of our airborne aircraft. I again explained the operations and function of each of the TOCs and how each TOC interacted with the two remaining Direct Air Support Centers (DASCs). The problem became evident as he tried to relate FAC operations to airlift operations. He really did not understand just what we did as FACs. His frame of reference for flight operations came from his background of airlift flight operations.

Military Airlift Command (MAC) Wings have total control over all of their flight operations. This includes weekly schedules, monthly schedules and even three-month schedules for their aircraft and crews. These schedules consist of assigned crews to specific aircraft for specific missions with specific destinations and return times and dates. He then informed me that since we were now assigned to their Wing, I was going to have to provide him with a weekly schedule, a monthly schedule and even a three month schedule for all of our flight operations. In addition, he also informed me that I would have to meet these schedules within 3 percent for weekly schedules, 5 percent for monthly schedules and 10 percent for the three month schedules. He said that I would have to explain any deviations from these percentages at each of the weekly scheduling meetings that I would now have to attend. He also informed me that the proper uniform for these meetings was the 1505 uniform. I was not sure I even had a 1505 uniform with me. I had lived in a flight suit for the last six months.

I knew right then he probably would not like the flight suit that I had converted into short-sleeved flight suit. When flying O-2As, we did not consider that our greatest concern was protecting our arms from a possible fire. We definitely had other greater concerns and the O-2A did not have air conditioning except by opening the windows.

I respectively tried to explain to him that I really could not provide any type of daily schedule let alone a weekly, monthly and a three month schedule. We flew when required to areas where "Charlie" was acting up, and there was no way for me to know our enemy's schedule. I said that the enemy essentially controls our schedule. I did explain that I could provide a daily schedule for the pre-planned air strikes that we knew about when we got the daily FRAG Order from 7th Air Force each night at 2200 hours (10:00 P.M.). My explanation did not go over very well with the Colonel. I could tell that since I was only a Captain and he was a full Colonel, this was going to be a real problem for me.

As an attempt to try and educate him on FAC operations and provide him with more information, I suggested to him that I would be happy to take him around to our FOLs to show him our daily operations. I hoped that this would explain why what he was asking for was going to be extremely difficult for FAC operations.

He agreed and we set a date for the following week for him to come to the Squadron for his orientation tour into the FAC world of combat flight operations.

A good thing happened, at least for me, before this scheduled day arrived. "Charlie" decided to launch a rocket attack against Tan Son Nhut AB around 1000 hours (10:00 A.M.) one morning. I immediately launched a FAC to address the situation. My FAC took off and was able to locate the area from which the rockets were being launched. The FAC requested air support from the DASC and two A-37s were scrambled out of Bien Hoa, which was located approximately 20 miles from Saigon. They arrived, and the FAC put them in on the target to address the problem. The air strike silenced the rockets and the FAC returned to Tan Son Nhut.

It was fortunate that this happened since this was not a scheduled flight, and the air strike was visible from Tan Son Nhut AB. The Colonel and other members of the Airlift Wing got a chance first hand to observe just what FACs do. Tell me where "Charlie" will act up, and I will tell you where I will be. This incident definitely helped me explain to the Colonel just why I could not meet his rigid scheduling requirements.

The Colonel arrived early on the morning of our scheduled flight to visit our FOLs. The weather was overcast at about 1500 feet with the cloud deck being about 500 to 1000 feet thick. I got all of his information and we went over to Base Operations to file our flight plan. We normally did not go to Base Operations since we would file our FAC flight plans from our TOC; however, since this was a full Colonel and this was his Wing, I was trying to comply with what he was used to when flying.

I briefed him that we would first go to Ban Me Thuot FOL, then Pleiku, followed by Nha Trang, Bien Hoa and then return to Tan Son Nhut. I figured we could visit four out of the six FOLs on this flight.

I chose Ban Me Thuot and Pleiku as the first two FOLs to visit since the Army's Special Forces (SF) were operating out of Ben Me Thuot, and the FACs at Pleiku were supporting MRII operations which at times still had active combat operations on going. These two FOLs were in the Central Highlands, which was known not to be friendly.

My first mistake was trying to comply with his expectations with a filed flight plan at Base Operations. I filed a visual flight plan (VFR) instead of an instrument flight plan (IFR). All airlift flight plans are IFR only.

He asked me why I did not file an IFR flight plan since we had an overcast cloud deck. I responded tactfully and said VFR is what we should use because Ban Me Thuot did not have a control tower or any active radar control capability. In addition, many of the navigational aids in that area were unreliable.

Ban Me Thuot was not a major base and only had a relatively short asphalt runway with no control tower. It also only had PSP taxiways that went inside a large bunker area where the SFs lived. I conveyed to him that essentially "Charlie" owned the runway at night, since it was outside the bunker area, and we owned it during the day. In addition, the TACAN Navigational Aid was located outside the bunker area close to the runway and was not always reliable.

He then asked me how we were going to get to Ban Me Thuot. I said that we would takeoff, climb-out above the clouds and head north. Ban Me Thuot was located approximately 60 miles north and approximately 20 miles from the Cambodian border. The Army's SFs were operating out of Ban Me Thuot and still were very active.

FAC missions in support of these SFs consisted of FACs accompanying the insertion missions for Long Range Reconnaissance Patrols (LRRPs) in areas like Cambodia and Laos. Each LRRP team would consist of five to seven SF members who would be inserted into areas of interest for intelligence gathering purposes.

The insertion team usually consisted of two Huey helicopters carrying the team along with an O-2A FAC. The Hueys would fly to the insertion point and allow the SFs members to deplane from a hover without actually landing. After the team was inserted, the Hueys and the FAC would move away from the insertion point and loiter for at least an hour. If the insertion was clean and we did not receive any radio transmissions from the team that the insertion had been compromised, we would return to base.

After five or six days, we had a schedule time and location to extract the team. The pickup point was always different from the insertion point. On the missions that I flew in support of pickup, we never got all of the team members back. These SFs members were a totally different breed.

During any briefings with these members, I learned that you do not look at them or talk to them unless they started the conversation. These team members only trusted each other and did not like to engage with outsiders.

The missions that these SF members went on were very dangerous. These individuals essentially were reduced to a basic survival level while doing their duty. These soldiers were really no longer the normal Army soldier but were killing machines for their own survival while in the field. Some of these SF members did have necklaces of ears that they had cutoff of enemy soldiers that they had encountered. One even had a human skull on his dresser in his room with a candle on top. Many of these SF soldiers became homeless upon return to the States because they just could not assimilate back into normal society.

For those of you who understand Maslow's hierarchy of needs, these soldiers had regressed from the higher levels, like self-actualization, down Maslow's pyramid of needs to the basic level of survival. Upon their return to the States, many just were not able to recover from their war experiences. War completely changed many of these soldiers.

After we had been flying for about 45 minutes, the Colonel asked me if I knew where we were. The cloud deck was still essentially overcast with only a few breaks in it. I assured him I knew where we were, and I told him that if the clouds were not present, he would be able to see a lake out of his side window. As luck would have it for me, a small break in the clouds was present on his side and sure enough he saw the lake.

Since Ban Me Thuot was a small FOL, the runway was short and not very wide. There also was no control tower present to control aircraft traffic. Ban Me Thuot really did not have much aircraft traffic except for the SFs helicopters and us FACs. In fact, the Army helicopters usually landed inside the bunker.

When I told the Colonel that Ban Me Thuot did not have a control tower, he asked how we were going to get below the cloud deck to land. Ban Me Thuot, was located in a valley surrounded by mountains. I told him that when we arrived overhead, we would let down in circles until we broke out below the clouds. We would then make a low pass over the runway to clear it and then land. I then contacted our TOC at Ban Me Thuot and informed them when we would get there. The Colonel seemed to be a little uneasy about how we were going to get down and land.

When we arrived over Ban Me Thuot, the mountains were sticking up through the clouds. I began our descent down into what looked like a bowl of clouds. When we broke out under the clouds right over the runway, I could not tell whether he thought I was just lucky or if I actually knew what I was doing.

To attempt to try and smooth over our potential issues, I asked him if he would like to try his hand at the landing. He replied, "Yes." I then suggested that we make a low pass over the runway to clear it of anything that should not be there. I told him that since there was no control tower we were on our own.

As we were making the low pass over the runway to clear it, we discovered that an A-1E aircraft had crashed the night before on landing. The A-1E was lying on the overrun at the end of the runway upside down. The Colonel then decided that maybe I should make the landing.

We landed over the upside down A-1E aircraft and taxied to the end of the short runway. There were only two taxiway exits from the runway with one being located at each end. These taxiways consisted of PSP panels that led into the SFs camp which was located behind six foot dirt bunker walls that surrounded the base. I asked the Colonel if he would watch his side wing tip for me as I watched my wing tip since there was only about a foot of clearance between the bunker walls and our aircraft as we taxied in. As we taxied in, I suggested to the Colonel that he not talk to any of the SFs as they really did not trust anyone that they did not know. I told him I would show him around our TOC where we could talk to our folks and then we would be on our way.

As we got out of the aircraft, the Colonel did see some of the SFs members walking nearby. These soldiers all had beards, long hair and uniforms that once were regulation uniforms. Many modifications, however, had been made to these uniforms and many different combinations were in evidence by these individuals. Upon seeing these individuals, the Colonel seemed to increase his pace while we walked to our TOC.

Once inside the TOC, the FACs at this location began to brief the Colonel on their FAC operations. The Colonel seemed to be uneasy about being there at Ban Me Thuot and really did not seem to be listening. I asked if he would like to see the living quarters of our FACs. He responded with a "No, that's all right," and suggested that we proceed on with the orientation tour.

We returned to the aircraft at a rapid pace, started up and taxied out to the runway for takeoff. Since there was no control tower, we just taxied out and took off. The Colonel did not say a word as we left Ban Me Thuot. I could tell that he was happy to be out of there.

It was still overcast when we took off so I started circling the field as we climbed-out to avoid the mountains. After we broke out on top of the clouds, we headed north toward Pleiku and the Central Highlands of Vietnam.

Pleiku was located in an area that was still somewhat of a "hot" area. Approximately 20 years earlier, the North had infiltrated the area with North Vietnam sympathizers to marry the local inhabitants and assimilate into the population. This had been very effective, and the Viet Cong (VC) were very strong and active in this area.

Pleiku was a much larger base and did have a control tower and operated much like a normal Air Force Base. The weather was also much better around Pleiku so our VFR arrival was no problem. We contacted the tower for landing instructions and proceeded to the traffic pattern for landing.

Pleiku had a normal length runway that could handle any size aircraft. They also had concrete taxiways and large covered aircraft parking areas for all types of aircraft. As we turned off the runway onto one of the many taxiways, we noticed that a C-141 was off of the taxiway and stuck in the muddy area between two taxiways. Ground control informed us to use caution to avoid the C-141.

Seeing one of his types of airlift aircraft stuck in the mud got the Colonel's attention. We did not know why this aircraft was located where it was, but it was obvious that no one at the present time was working on getting the aircraft out of the mud. I was sure that there was some story behind this aircraft, but we did not know what it was. This seemed to really get the Colonel's attention.

We taxied into our FAC parking area and shut down. We were met by a couple of the FACs that were assigned at this location who took us into the TOC. The Colonel listened as the FACs described their operations in support of the combat operations being conducted in their area of responsibility. We were supporting ARVN troops as well as some South Korean troops that were still present in MRII. After the briefing, the Colonel was ready to leave.

After we got airborne, I informed the Colonel that we would now visit the Nha Trang FOL. He said that he did not think that was necessary and that he would prefer to return to Tan Son Nhut. I kind of had the feeling that he had seen enough and did not want to know any more about FACs.

After takeoff and his comments, instead of heading east toward Nha Trang, I turned south for return to Tan Son Nhut. The weather had greatly improved, and we were flying above a scattered deck of clouds. We could easily see the tops of the mountains that we were flying over. The Colonel was looking out his side of the aircraft at these mountains when he asked me what the small indentations were along the tops of the mountains. I rolled the aircraft over to his side to see what he was looking at. What he was seeing were 50-caliber gun pits that the VC had constructed along the tops of these mountains. I told him not to worry as there were no guns currently present in these gun pits.

I then relayed to him how the North had infiltrated this part of South Vietnam with Northern sympathizers and that this was not a very friendly area. I told him that this was an area where we had lost a FAC, and when we found his body, his Geneva Convention card had been nailed to his forehead. He did not say anymore the rest of the way back to Tan Son Nhut.

Upon landing at Tan Son Nhut, we got out of the aircraft. The Colonel got out and thanked me for the tour. He then paused and said, "I think I have a better perspective of what your FACs do now. I also saw how you addressed the rocket attack last week, and now I can see you really do not have a lot to say about your schedule." He then informed me that even though we were assigned to the Airlift Wing, I should do whatever I needed to do to complete our mission. He also said that I did not have to submit any schedules to him nor did I have to attend his weekly scheduling meetings. He also said that he would provide any support that I needed to do our job. This was very good news to me since I did not have any 1505 uniforms, and I knew I could not meet his requirements for scheduling.

For the next couple of months things were generally quiet as far as the war was going in South Vietnam. Things in Cambodia, however, were not as quiet. There seemed to be an increase in enemy activity in Cambodia. We were providing tactical air support to the Cambodians who were fighting the Communist Khmer Rouge.

Our area of responsibility in Cambodia extended from the South Vietnamese border west into Cambodia to the Mekong River to Phnon Penh, then north along the Mekong River to the Laos border. We flew sorties that provided tactical air support for Cambodian "troops-in-contact" (TIC) as well as reconnaissance flights looking for "Charlie" and/or truck traffic. We also provided combat patrol missions in support of the U.S. Navy who would transport supplies to Phnon Penh via the Mekong River. If any of these river convoys were attacked, we would call in air strikes to engage the enemy. Getting clearance to expend ordinance in Cambodia was even harder than in South Vietnam.

I do remember a couple of my missions in Cambodia that stood out from the others. On one mission when I was flying a reconnaissance mission along the Mekong River, I started taking 23-mm fire from a Pagoda. This Pagoda was located in a small village that consisted of 30 to 40 structures. I reported the 23-mm activity to our TOC and proceeded away from the area. I gave the TOC the coordinates of the 23-mm because I knew I could not get clearance because of our Rules of Engagement (ROE).

Our ROE stated that we could not engage the enemy in any Pagoda nor could we strike any village with 25 or more structures. Since this target failed both of these rules, I decided to get out of the area. 23-mm guns are not fun targets to attack since they can fire up to 700 rounds per minute. In addition, we flew well within their engagement envelope. Quad 23-mm (four barrels) guns were a very serious problem for us. Thankfully, this one was only a single barrel.

I was leaving the area to look elsewhere when our TOC contacted me and asked if I wanted clearance for the target. I again reminded them that the gun was firing out of a Pagoda window that was located in a village that had more than 25 structures. I am sure that "Charlie" also knew our ROE and, as such, he felt very safe firing at me. The TOC responded with a "Yes," that they understood, but do I want clearance?" I responded by saying, "If you can get clearance, then, yes, by all means, I want it." Five minutes later I received clearance and was told fighters were on the way.

When the fighters checked in, I briefed them on the target and told them to watch their tails since this was an active 23-mm site. After 10 minutes, the Pagoda was destroyed along with several nearby structures. There also were secondary explosions going off as we must have hit an ammo storage facility as well. That was a very good mission, but I never understood how I got clearance and I never asked.

A second mission in Cambodia that stood out occurred when I was again on a reconnaissance mission. I spotted a convoy of trucks on a road that was heading into the Chup Rubber Plantation. The Chup Rubber Plantation was a very large plantation that supplied raw materials to the Michelin Tire Company. The Chup Rubber Plantation was also one of those targets that went against our ROE.

The truck convoy entered the Plantation from the north and then disappeared. This peaked my interest. I made a number of passes over the Plantation attempting to locate the truck convoy. From the air and the view that I had, I knew there was no way that they could conceal all of those trucks above ground so fast. I had to assume the trucks had entered a large underground bunker since they were nowhere to be seen. Since the Chup Plantation was on the "no strike" list, I called in what I had observed to our TOC and proceeded on my way.

About 10 minutes later the TOC contacted me and asked me if I wanted clearance. I responded with "You know that this is the Chup Plantation, don't you?" They confirmed the location as the Plantation and again asked if I wanted clearance. I replied, "If you can get clearance, they by all means I want it." I knew approximately where the trucks had disappeared after they entered the Plantation. Approximately 20 minutes later the TOC indicated that we had clearance. I then requested any air available.

A set of F-4s were the first to check-in. I briefed them on the target and told them we were going after an underground bunker that was a truck park. I marked the target area and requested that each fighter lay down a string of their 500-lb bombs in the target area with a separation of 100 to 200 feet between the strings. After the F-4s had expended their ordinance, I made a pass over the target area to assess the damage.

We were right on target. I noted large tree logs, as much as 24 inches in diameter, that were splintered and sticking out of the ground. These logs were obviously the roof support structure of the underground bunker. I immediately contacted our TOC and requested more fighters.

I put in two more sets of F-4s that afternoon. By the second set of fighters, we had done enough damage to the bunker roof structure that the bombs from the last set of fighters actually penetrated the bunker itself. There were numerous secondary explosions going off with debris being thrown up to 2000 feet in the air. In addition, there were large fires burning and there was smoke everywhere. We had hit a major stockpile of ammunition in an underground storage bunker in this Plantation.

I am sure that "Charlie" felt his supplies would be safe in this underground bunker since it was the Chup Plantation. I am also sure that he knew our ROE or he would not have been so bold as to be moving these supplies in broad daylight.

Later that day an RF-4C reconnaissance bird made a pass over the target to assess the damage. They reported that secondaries were still going off. I am not sure just who actually gave us clearance to strike this target, but we sure did hit the jackpot on that day.

I never again got clearance to hit the Chup Plantation. I also found out later that our government reimbursed the owners of the Plantation $300 for each tree that was destroyed. To me it was worth it.

I also really enjoyed providing tactical air support to the Cambodians. They are a very proud people and really appreciated all of the help that we gave them. I felt that they appreciated our support much more than the South Vietnamese did. I also felt that the Cambodians were better fighters than the South Vietnamese. They seemed to want to defend their country more than a lot of South Vietnamese soldiers did.

Working with Cambodians was a real pleasure. They very seldom would request air support on their own even when they were in a "troops-in-contact" TIC situation. The way we would support these Cambodian troops would be for us to contact them. When we knew where their troops were located and who was in charge, we would fly over their location and contact the ground commander. We had to be very tactful on how we approached them.

After initial radio contact, the communications had to be very respectful and gracious so as not to insult them or suggest that they needed our help. We normally would start the radio conversation out with a comment like, "Roses to your family." They would respond with, "and roses to yours." We would then comment on the weather and indulge in other small talk for a few minutes.

I remember on one mission, I encountered a TIC situation, but I had to go through the ritual as described above. After our initial salutations and small talk, I asked the Commander, "Since I am here, is there anything I can do for you and your troops?" I could see that they really needed help but I could not suggest that he could not take care of himself or his troops. His response was, "Yes, since you are here, we could use some tactical air support if you happen to have some available." I immediately contacted our TOC and two AT-37s were scrambled out of Bien Hoa. The Air Force kept these AT-37s on alert at Bien Hoa primarily for air support to Cambodia. After I put in the AT-37s, the TIC condition was broken. The Commander and I exchanged pleasantries, and I left the area with a very good feeling about the mission.

During the early part of 1972, President Nixon's Program of Vietnamization continued. Our combat troops were being continually reduced as more and more areas were turned over to the Vietnamese. Combat operations for the U.S. were winding down. FAC flight operations during January and February of 1972 were becoming very routine, and not much action was going on. This turned out to be the lull before the storm.

CHAPTER EIGHT

The Spring Offensive of 1972

ON MARCH 30, 1972, everything changed as far as the war was concerned. North Vietnam launched a major offensive into MRI. Troops came across the DMZ south and from the west out of Laos into Quang Tri Province. They were very aggressive and caught the South Vietnamese forces by surprise. Since MRI was under the control of the South Vietnamese forces, they redeployed some of their units from MRIII to assist MRI. The 20th TASS was still present at Da Nang and was able to provide tactical air support from the U.S. units that were still available. The U.S. Navy still had assets off of the coast, and the Air Force still had fighter support based in Thailand. The Army and Marines also still had advisors with the ARVN units so they still had access to U.S. tactical air support.

It was not long before we found out that the North had planned a three-prong attack on the South. A second front was opened up by the North in MRII just west of Pleiku near Kontum in the Central Highlands. North Vietnamese troops entered South Vietnam from the southern Laos/northern Cambodian border area through the An Kay Pass. These troops were heading east and seemed to have as a goal of cutting the South in half. If they could get to the coast, they would have access to a seaport from which they could resupply their forces by sea and not have to depend so heavily on truck traffic and the Ho Chi Minh Trail system of resupply. If successful, this would be a tremendous boost to their combat capability.

In MRIII, the third prong of their offensive was initiated out of Cambodia near the South Vietnamese village of Loc Ninh. Loc Ninh was located about 100 miles north of Saigon and had a major road that ran south through An Loc to Saigon.

It soon became obvious that this three-prong attack was a major offensive by the North to take over South Vietnam. I think the North felt that since the U.S. was withdrawing its forces rapidly, they felt they could now achieve their goal of taking over the South and embarrass the United States as we withdrew. This three-prong attack put a significant strain on the South Vietnamese forces as well as the remaining U.S forces.

Two of the three-prongs of the attack by the North were in MRII and MRIII–my areas of responsibilities for tactical air support. My life as well as all of the FACs in MRII and MRIII changed significantly and in a very short period of time. All of the FOLs in MRII and MRIII became very active with a "hot" war again.

It was a very good thing that each FOL operated essentially autonomously. It would have been virtually impossible to control all of combat operations if operational control had been attempted by a central location. It was also good that the Airlift Wing decided to leave us alone and just provide any support that we needed. They now truly understood why we could not meet their scheduling requirements of a centralized command organizational structure. We were once again in a "hot" war now on three specific fronts with the ground combat load being carried by South Vietnamese forces.

A number of other things became clear by this offensive outbreak by the North. It was now evident that the missions I had flown in Cambodia against the 23-mm and the Chup Plantation were a prelude to this offensive. This was evidenced by the large number of secondary explosions that occurred with the targeting of these two locations. Both targets were obviously large ammunition stockpiles that were being prepared for this offensive.

In addition, the mission where I encountered the 23-mm in Cambodia also suggested that they were preparing for something big. We normally only encountered 23-mm when something big was going to happen and normally not this far south. It was all becoming very clear as to the intent of the North.

The FACs in MRII had also been reporting 37-mm guns present in that region. This was the farthest south that we had encountered this size of antiaircraft guns. We also were beginning to encounter SA-7 shoulder launched missiles. With all of this increase in activity, it should have been obvious that the North was indeed planning something big. The amount and size of these weapons suggested that regular NVA forces were also present. The VC had not tried to engage us with anything larger than 50-caliber guns other than an occasional 23-mm gun. The presence of 37-mm and SA-7 missiles represented a whole new level of battle for us.

The 37-mm did not bother us too much since these rounds were airburst rounds that usually went off well above our altitude, and they could only fire seven rounds per minute. The SA-7, however, did represent a significant threat to FACs.

Our flight operations placed us well within the SA-7 missiles envelope of engagement. In addition, the O-2A did not have any defensive system to defend itself against these shoulder launched missiles. These missiles had an infrared guidance system and, once launched, they did not need radar or any other type of ground control. Once launched, the infrared sensor would start searching for a heat source as its target. The O-2A, with its two engines that were heat sources, were prime targets for SA-7 missiles.

After some losses to SA-7 missiles, we did, however, develop a defensive maneuver against these missiles, if we were lucky enough to see the missile launched. Of the two engines on the O-2A, the rear engine ran about 200 to 300 degrees hotter than the front engine due to less cooling airflow. The enemy also usually launched the missile after we passed by them. This, unfortunately, exposed the highest heat source for the missile to lock on to.

The defensive maneuver that we developed was almost a suicide maneuver. If we did see the missile launch, we would wait until it locked on to one of our engines. We could tell when it locked on because as it left the launcher, it had a corkscrew flight path of smoke while in the search mode. When it locked on, it would kick toward us, and the smoke trail became steady as the missile's sensor tracked its target.

The maneuver that we would perform was to watch the corkscrew flight path until it locked on to one of our engines. This was evidenced by a sharp kick into us followed by a steady smoke stream trail heading toward us. We would then make a hard break into the missile just like we were playing chicken with the missile. By doing this maneuver, we hoped that the missile had locked onto the rear engine since it ran hotter than the front engine. By breaking into the missile, we were shielding the rear engine from the missile with the front engine and the cockpit. If the missile had locked onto the rear engine, it would break lock and start searching again as evidenced by a corkscrew flight path.

During this maneuver, it was helpful if the weather conditions consisted of a scattered to broken deck of clouds above us. Many times the missile's sensor would lock on to the heat of the sun coming through a break in the clouds. The missile seemed to like the sun's heat more than the heat of our front engine. This maneuver worked about 90 percent of the time, if we were lucky enough to see the missile launched. At our altitude, the time of flight for these missiles before contact was five to seven seconds.

One of my young Lieutenant FACs encountered a quad launch of SA-7 missiles on one of his missions. It was the first time that we had encountered a quad launch of SA-7 missiles. He reported that the launch of the four missiles was almost simultaneously, which was lucky for him. In addition, all four missiles seemed to lock on at nearly the same time. When they did lock on, the Lieutenant (LT) performed the hard break into all four of the missiles. Fortunately for him, all four missiles had locked onto the rear engine and all four missiles did break lock. Two of the missiles passed over the top of his right wing while the other two missiles passed under his right wing.

He immediately returned to base having been totally shaken by this near miss. As he was telling me about his encounter, he was still very white and his hands were shaking. God was definitely watching over him that day. After a couple of days off from flying, he once again rejoined the battle.

With the North initiating this three-prong attack and two of the three prongs located in MRII and MRIII, the 21ˢᵗ TASS became extremely busy. We had to provide 24-hour airborne coverage in both these regions and still support Cambodia. Our FACs began flying two or three missions every day in support of combat operations to stop the North. This was a very intensive time period for all of us.

As Squadron Operations Officer, I decided that I would fly two missions each night and then run the Squadron during the day to keep up the required support to the FOLs. With the tremendous increase in flying time, the time between required periodic inspections of the aircraft was greatly reduced. This caused me to have to move aircraft around to keep safe aircraft in the field for use. I also had to move FACs around to different FOLs as required to meet the threat.

For the next 30 days (the month of April 1972), I would sleep two hours between 1800 – 2000 hours (6:00 P.M. to 8:00 P.M.) and then fly two night missions. Upon completion of these two missions, I would sleep between 0600-0800 hours (6:00 A.M. to 8:00 A.M.) and then run the Squadron until 1800 hours when I would repeat the cycle. During this time period, I lost about 25 pounds because I did not really have time to eat properly. I ate on the run whatever I could get a hold of that was food and fast. I do not recommend this as a good diet plan.

The offensive in MRIII started on the 6ᵗʰ of April 1972 at the village of Loc Ninh. Loc Ninh is located approximately 100 miles north of Saigon near the Cambodian border. I was the FAC on station when the offensive started. An American Marine advisor with the ARVN unit at this location requested immediate air support when a large North Vietnamese Army unit attacked their fire base. There was no doubt that this was going to be a major offensive. The ground action reminded me totally of my mission is support of FB Sarge when I was assigned to Phu Bai in support of the 101st Airborne. The only difference this time was that these were ARVN forces and a Marine Major advisor on the ground instead of the 101st Airborne. I flew two missions in support of this FB that night, which essentially started my intense schedule for the next 30 days.

The next day, the ARVN unit at this FB had to retreat to An Loc which was approximately 20 miles south of Loc Ninh. The FB at An Loc was larger than the one at Loc Ninh so the ARVN decided it was better to defend at An Loc. The battle for An Loc was fierce and brutal. During the day I had three airborne FACs on station to support this FB. At night I had two airborne FACs on station to provide support.

Our battle plan for defense of this FB consisted of the three FACs during the day. One FAC would act as a gatekeeper and keep track of the fighter aircraft as they checked in. The fighters were stacked in a holding pattern until one of the other two FACs were ready to put them in. This allowed us to have constant air support during the day.

At night we could handle the air strikes with two FACs since the number of air strikes was reduced but still more than one FAC could handle. I used my more experienced FACs to direct the fighters and the younger less experienced FACs to act as the gatekeepers. This worked pretty well.

The battle for An Loc went on for approximately 10 days. The North was taking heavy losses but were resupplying and reinforcing their units daily. For us to resupply the ARVN forces at An Loc, we used C-130 aircraft, which would air drop supplies for the FB. This was a very dangerous mission for our C-130 aircrews as they had to fly very low and airdrop the supplies next to the FB. We lost one C-130 during one of these resupply missions. We could not get any helicopters into the area to rescue the crew due to ground fire. The ARVN from the FB, however, were able to recover the C-130 aircrew. It then took three days before we could get the crew back to Saigon. The pilot of the C-130 had been a fellow IP with me at Laughlin AFB, and I had seen him the day before.

With the schedule that we were flying we were rapidly flying our aircraft out of flying time between inspections. Our maintenance personnel worked their tails off trying to keep our aircraft flying. We were forced to start waving these required inspections because we could not go without the aircraft. We also were not able to comply with the Air Force's crew rest policy for our pilots.

I remember one night when I returned from my first mission of the night, I called into our maintenance to let them know I needed to return to An Loc as soon as possible. I was informed that we did not have any serviceable aircraft ready to fly. I inquired what was wrong with each of the aircraft that had been grounded from flight. Most of the aircraft had engine problems which required repair or engine replacement. I asked if any of the grounded aircraft had two good engines. The response was "Yes, but that aircraft was missing the attitude indicator." This instrument is the primary flight control instrument for all aircraft. The attitude indicator is the flight instrument that pilots use to maintain level flight in weather or at night. I told them to prepare that aircraft for flight with fuel and rockets because I did not have a choice. I had to return to An Loc since the lives of those on the ground were depending upon me to give them the tactical air support they needed. I knew the weather was good between Tan Son Nhut and An Loc and I would not be looking inside the aircraft to maintain control.

This probably was not the smartest decision on my part but the situation was critical for those on the ground. I also was not going to ask anyone else to go. In fact, I would have gotten after anyone who would do such a thing. In war, however, you do whatever is required to save lives.

It was also during this time period that I developed my opinion about why I felt the Cambodians were better soldiers and cared more for their country than many of the South Vietnamese did. A number of times I had to load my own rockets when our munitions personnel were overloaded by the schedule and had not been able to get all of the aircraft ready for flight. We had hired some South Vietnamese to help with rocket building and loading of rockets onto the aircraft. This helped considerably until the monsoon season started.

When the monsoon season started and we would have a shower, many of these Vietnamese workers would not show up to build and load rockets. They were more interested in gathering "rice bugs" from drainage ditches when it rained enough to have water running in these ditches, which ran alongside the roads on Base. Evidently these "rice bugs" were a delicacy and I was surprised at the number of Vietnamese that would be collecting these bugs after each rain. They would bite the heads off of these bugs and then suck out the insides. It appeared these bugs meant more than saving their country.

I was also the FAC on station when An Loc fell. I was working with the Marine Advisor when he informed me that the NVA had breached their perimeter. He said that "Charlie" was coming through the wire and that I should hit his position. I asked him if he was sure. He responded with a "Yes, that he would not be able to tell if the ordinance hitting him was ours or theirs." He then said, "Just hit my position."

His next radio transmission really hit home. He said, "My wife is in Bangkok. Tell her I love her." The radio then went completely silent. That was one of the worst moments during my entire time in Vietnam. I will never forget that feeling of helplessness that I felt at that moment.

With the radio being silent, I called our TOC to see what they wanted with the remaining air strikes that were waiting. They responded by saying hit the FB and then take out the northern six blocks of An Loc. They said there are no friendlies left in An Loc.

I completed the mission and returned to Tan Son Nhut. I was devastated by the events of the night. I knew the situation was tough on the ground, but this mission really brought it home.

For the next two weeks our efforts were to try and stop the advance by the North down the highway from An Loc to Saigon. They were now only about 70 miles north of Saigon. The South Vietnamese deployed their troops along this highway north of Saigon in their attempt to stop the advance. We continued FAC coverage day and night to assist in their efforts.

It was three days after my last contact with the Marine Major at An Loc when I got the best news that I could have received. I was at my desk working on the schedule when one of my Sergeants came into me and said, "Someone is looking for Chico 10." Chico 10 was my call sign when I was flying out of Tan Son Nhut. I responded with "Bring him in." In walked a Marine Major wearing a poncho and covered with red dirt and sporting a 10-day growth of beard. He had red hair, which almost matched the red dirt.

When he saw me, he asked, "Are you Chico 10?" I responded with a "Yes." He rushed over to me and said, "I want to shake your hand. I am the guy whose life you saved at An Loc." He said when the air strikes began to hit his position, he was able to escape to the south from the FB. He indicated that it had taken him three days to "escape and evade Charlie" and he wanted to thank me for saving his life. He grabbed my hand and began shaking it with great enthusiasm. I smiled at him and said, "You can tell your wife that you love her." He smiled back and at that moment a bond forever had been established between him and me. I do not remember his name, but I am sure he remembers me just as well as I remember him, without names—just brothers in combat.

Before he left, I asked him if I could get him anything. He responded with, "I could use some dry socks." I had him sit down and told him I would be right back. An Army supply unit was next door to our flight operations building.

I hurried over to the supply warehouse and told the Army Sergeant what I needed and for whom. Without any hesitation, he gave me a dozen new pair of socks for the Major. I returned to my office and gave the Major the socks. His face lit up like a little child getting a Christmas present from Santa Claus. He took the socks, stuffed them in his backpack, shook my hand again and thanked me once more and left. He said that he was on his way to Bien Hoa where he would be assigned to another ARVN unit and go back up the road to stop "Charlie." I never met a Marine I did not like. They are awesome!

My night missions over An Loc left me ingrained with an extremely strong dislike of two very normal things. I now do not like any fireworks displays even on the 4th of July. Fireworks remind me of 23-and 37-mm guns.

My second dislike is of those large search lights that car dealerships and others place outside their businesses to draw attention to their location. "Charlie" was using these types of search lights trying to find us over An Loc. We flew these night missions without exterior lights so they would try to locate us with these search lights. They knew if they could shoot down the FAC, air strikes were less likely to cause them problems, especially at night. To this day whenever I see fireworks or those search lights, I am again immediately in an O-2A over An Loc.

It was also during this time period that I encountered a very difficult situation with one of the new FACs who had just arrived. The individual was a young Captain who had been flying C-141s after completion of pilot training, and this was his first assignment after that.

After he processed into the Squadron, I was talking with him and explaining how we operated and what would be required of him. He was very quiet and did not respond much in any way. I told him I would schedule him with an experienced pilot to check him out. I told him we would start his checkout the next day. He then asked me if there was any way that I could get him assigned to Korea. We did have O-2A FACs in Korea, but he had been assigned to Vietnam. In fact, some of the Korean FACs had been sent TDY (Temporary Duty) to Vietnam when we were short on

FACs. His question raised a significant concern since I had never been asked that question before. I told him that I did not have the authority or capability to get him transferred to Korea. I told him to check the schedule in the morning as I would have him scheduled to fly.

The next morning this Captain came into Operations to check the schedule. I had scheduled him for a flight in support of Cambodia. I did not want to start him out on a mission in a real "hot" area. After looking at the schedule, he asked if he could talk to me. I, of course, agreed and took him over to my desk. He sat down but did not say anything. After a couple of minutes of silence with him looking down at the floor, I asked him what he wanted to talk about. In a very soft voice, without looking up from the floor, he responded by saying, "I can't fly." This, of course, did not go over well with me since all of my other pilots were flying their tails off. I responded by saying, "You have to fly." I then went on to say that I just could not have him around the Squadron and not be flying missions. I then told him that if he was refusing to fly combat, I would have to ship him back to the States with a recommendation for a Court Martial. I then asked him why he could not fly. He responded with, "I just can't." I told him we were not leaving my office until he told me why he could not fly combat. He just kept saying, "I just can't fly combat missions."

After we sat there for about 10 to 15 minutes with him not saying anything, I asked him if he could not tell me why he couldn't fly combat missions, could he write it down for me? He responded with a "Yes," he thought he could do that. I got him a tablet and pencil and took him over to an empty desk that was not far from me. I then went back to my desk to work and left him alone to write.

I kept watching him from my desk, but I did not see much writing going on. After about an hour and a half, I got up and went over to the desk he was at and asked him if he was finished. He did not look up, but he did respond with a "Yes." He handed me the tablet that I had given him to write on. On the tablet were four words. They were "I can't take life."

I took him back over to my desk to try to get more information from him. I asked him to explain what he meant by his written statement. He paused and then said, "Well, if I put in an air strike which kills the bad guys, I take life. If I don't put in the air strike and the bad guys kill the good guys, then I am responsible for their loss of life. Either way I would be responsible for the loss of life." I then asked him if this was part of his religious beliefs. He responded with a "Yes." This caused me to pause since I now had a Captain who was refusing to fly combat due to his religious beliefs. This was definitely going to be a problem for me.

I then asked him why he even joined the Air Force in the first place if he felt this way. He paused and then said that he felt he had an obligation to serve his country. He went on to say that he felt that he could do that by flying airlift-type missions like those flown by C-141s. He said he never thought that he would be asked to fly combat missions, which would require him to make decisions on who lived or died. When he got the FAC assignment to O-2As, he asked to go to Korea since he knew that he would not be flying missions where this type of decision would be required. He did not know that some of the Korean O-2A FACs were coming to Vietnam on temporary duty. This caused me to pause because I truly felt he was telling me the truth.

I did not say anything for a while as I was trying to decide what I was going to do. I could not have him around the Squadron and not be flying. I then asked him how long he had been in the Air Force. He responded with a little over four years. I then asked him if he planned to stay in the Air Force for a career. He responded with a "No, he was not going to now since he now realized that he may be placed in similar situations in the future."

This gave me an idea. I did not have the authority to get him transferred to Korea, but I did have the capability to get him assigned to a non-flying job at the Direct Air Support Center (DASC). At the DASC his job would entail the coordination of getting fighters to FACs and obtaining the required clearances for targets. I asked him if he felt that he could complete this type of duty. He responded with he thought he could do that. He said that he felt by doing this type of job he would not directly be responsible for taking life. I then told him that if he went over to Personnel and requested a "date of separation" (DOS) from the Air Force upon rotation back to the States, I would get him assigned to the DASC. This tour in Vietnam would fulfill his obligation to the Air Force after pilot training. He said that he would do that.

The next day the Captain went to Personnel and obtained a DOS upon his return to the States. I then made arrangements to get him assigned to the DASC. The Captain departed and both of our problems had been solved. I had reached a solution that both of us could live with.

The ARVN were now beginning to slow the NVA advance to the south from An Loc in MRIII. The NVA advance in MRII was, however, still going strong. I had to shift some of our assets north to FOLs in MRII to help counter this advance by the North. I also began flying missions in MRII myself out of Tan Son Nhut. At times, I would fly two or three missions before returning to Saigon.

In MRII the North had broken through An Kay Pass and were heading east toward either Kontrum or Pleiku. The remaining South Korean forces that were still in-country were trying to retake the An Kay Pass. As I have previously noted, the South Korean forces were fierce soldiers and I really respected their efforts. I know why North Korea had not tried to take over South Korea after the fighting ceased in Korea back in the 1950s.

I gained this respect on one of these missions in MRII while supporting the South Koreans. The NVA had constructed concrete pill boxes in the mountains around An Kay Pass. The South Koreans were taking heavy losses from these pill boxes as they were attempting to take the Pass back. The South Koreans had requested our tactical air support to assist in neutralizing one of these pill boxes. We had 2000-lb bombs that could neutralize this threat. It was on one of these missions when some South Korean troops had been pinned down by heavy 50-caliber machine gun fire from one of these pill boxes when they requested our support.

I arrived on station and was briefed by the Korean Ground Commander on their situation. I knew it was going to be difficult to provide them the air support that they needed. Their position was such that they were very close to the target; and, due to the mountains, our fighters would have to overfly their position to attack the target. It was on the second set of fighters that I was directing when an incident occurred that all FACs dreaded—a "short round."

After the fighters checked in with me, I briefed them on the situation. I informed them that the target was a concrete pill box with active 50-caliber machine guns present. I also informed them of the restricted access to the target due to the mountains, which would require overflight of the friendlies (South Koreans).

I asked the fighters if they understood the situation and if they could handle it. They both responded that they understood and that they could comply. I marked the target and asked the South Koreans to pop smoke to identify their position to the fighters. Both fighters acknowledged they had the target location and the friendly location. We proceeded with the attack.

Lead rolled in and after I visually verified his location, I cleared him hot. Both of his 2000-lb bombs landed right on target. As two rolled in and he confirmed the target location and the friendly location, I cleared him hot. My worse fear came to fruition as one of his two 2000-lb bombs released prematurely and exploded directly on the friendly location. I directed both fighters to "safe them up and hold high and dry," while I checked with the South Koreans.

I contacted the Korean Ground Commander to see if the last bomb had been a problem. He responded with a "No" and to "keep dropping bombs." I asked him if he was sure and he responded with "We will lose more people if you don't get the pill box." We continued the attack, and put in a third set of fighters.

We were able to silence the pill box that day, but I know the "short round" had to have impacted the South Koreans. They were, however, happy with the results of the pill box and I never heard a word about the "short round." I gained even more respect for the toughness of the South Korean forces that day.

Over the next two or three days, the NVA were finally able to break completely through the An Kay Pass and head east. It soon became apparent that the North was heading to Kontum and not Pleiku. They must have felt it would be easier to take Kontum instead of Pleiku. The U.S. still had a significant present at Pleiku.

From Kontum they headed east toward Qui Nhon, which was on the east coast and could be a seaport for their resupply. Their order of battle seemed to be to approach Qui Nhon from the west and north. Some of the NVA forces were approaching the coast approximately 40 miles north of Qui Nhon and the small village of Bong Son. This was a small village along Highway 1 which runs along the entire coast of Vietnam. This Highway runs from Saigon north along the coast to the DMZ. At the DMZ, Highway 1 continues along North Vietnam's east coast to Hanoi. Highway 1 is the major highway for both North and South Vietnam.

The South Vietnamese forces were essentially slowing the North's advances in MRI and MRIII. In MRII, however, the North was still advancing. This area was not as populated as MRI and MRIII. As noted earlier, the North had infiltrated this area with their sympathizers 20 years earlier. This area had a lot of local support for the North's efforts. In addition, most of our U.S. Army forces in this area had been sent home as part of President Nixon's Vietnamization Program. That left this area more vulnerable than MRI and MRIII. The South Vietnamese forces in this area were less experienced, which resulted in less combat capability being available.

When the U.S. Army Forces in MRII left, an Army Lieutenant Colonel (LTC) who had just retired was selected to remain in-country to watch over combat activities in this region. His name was John Paul Vann. I am not sure who he worked for but it was most probably the Central Intelligence Agency (CIA). I suspect this since this man seemed to be very powerful and could get whatever support he needed for MRII.

I had two missions in MRII that were at the direct request of Mr. John Paul Vann. These two missions were against the NVA, and both were around the Bong Son area along Highway 1.

The first mission that Mr. Vann requested and I supported was when the NVA was approaching the eastern coast of South Vietnam at the village of Bong Son, which was north of Qui Nhon. The South Vietnamese forces that had been stationed at a FB near Bong Son had abandoned their FB and had fled south. When they left this FB, they did not disable their 105 Howitzers. Instead they left 12 of these Howitzers in perfect working condition for the NVA troops to acquire and use against the South.

Mr. Vann was able to get the Navy to provide him a fire support mission from the Battleship Missouri. This Battleship had been recommissioned and was stationed with the 7th Fleet off the east coast of Vietnam. Mr. Vann wanted a fire support mission from the Missouri and their 16-inch guns to destroy the abandoned 105 Howitzers at the FB. These 16-inch guns were capable of shooting an approximate 250-lb projectile for distances of 16 to 17 miles.

When I arrived on station, the weather was perfect; and I could clearly see the Missouri off of the coast to the east. Mr. Vann briefed me from his helicopter on the location of the Howitzers and said he wanted them all destroyed before the North got to this FB. I then contacted the Missouri's Fire Control Officer and briefed him on the target. I provided him with the target coordinates, and we proceeded with the fire mission. It was the first and only time that I directed Navy gunfire, and I was impressed!

The Fire Control Officer would give me "time of flight" for each round fired. The "time of flight" for these rounds from the Missouri's location to the target ranged from 55 to 58 seconds. The target was approximately 12 to 13 miles from the Missouri. The Fire Control Officer would call "Round Out," and I would observe a puff of white smoke from one of the 16-inch guns. I would start my clock and then observe the target for impact. I would then make any adjustments on the coordinates, and then we would repeat the process.

After approximately 45 minutes, we were able to destroy all of the Howitzers that had been abandoned. I was truly impressed just how far those 16-inch guns could shoot and just how accurate they were at that distance. I really enjoyed working with the Navy on this fire mission.

The second mission I flew in support of Mr. Vann occurred a couple of days later. The NVA had captured the village of Bong Son and were prepared to advance south toward Qui Nhon along Highway 1. When I arrived on station for this mission, Mr. Vann again briefed me from his helicopter on the target. He said that he no longer wanted the center span of the Bong Son River Bridge. This was a major bridge along Highway 1 and certainly would restrict the NVA from moving south with tanks or other heavy artillery weapons. Mr. Vann went on to say that he had F-4 aircraft inbound with laser-guided bombs to destroy the bridge.

I was totally impressed that Mr. Vann had been able to get a battleship for a fire support mission and now he was able to get laser-guided bombs. This was the only time I ever had laser-guided bombs on fighters that checked in with me. Laser-guided bombs were new and very expensive. Like I said, Mr. Vann must have had significant pull with someone very high up to get this type of resources for the support he needed.

When the F-4s checked in with me, they each had four laser-guided 500-lb bombs. I briefed them on the target as the center span of the river bridge along Highway 1 coming out of the village of Bong Son. The target was very obvious so I did not have to mark it.

I cleared lead in hot as I watched the bridge. Lead's bomb hit the center stripe of the center span and caused significant damage. I cleared two in to hit the remaining bridge structure of the center span. After the two bombs had impacted the target, the center span dropped into the river. Lead then asked where I wanted the remaining bombs.

I contacted Mr. Vann to see where he wanted the remaining bombs dropped. He responded by saying, "Drop the rest of the bombs on Bong Son Village. There are no friendlies remaining in Bong Son." The remaining six bombs were dropped on the village.

I was totally amazed at the amount of power Mr. Vann had in prosecuting the war in MRII during this offensive. That was the last time that I had any contact with Mr. Vann. I will say that in my opinion the decision to destroy the river bridge on Highway 1 by Mr. Vann was the critical decision that resulted in the stopping the 1972 Spring Offensive. By stopping the North from gaining a seaport at Qui Nhon, they were not able to effectively resupply their troops. Within a week, the level of hostilities in both MRII and MRIII began to subside. The only sad note was when I learned that Mr. Vann had been killed in a helicopter crash on June 9, 1972 near Kontum. He was a real hero in my opinion.

It was also during this time period when I was flying missions in MRII out of Saigon that I had my most hair-raising mission. I had flown three missions that day in MRII while checking out a new FAC. It was on our fourth mission, and we were returning to Tan Son Nhut at night.

We had taken off from Pleuki and had climbed up to an altitude of about 8500 feet and were heading south. As previously noted, MRII consisted primarily of mountains in the Central Highlands. In fact, the highest peak in South Vietnam is Ngoc Linh at 8524 feet and is north of Kontun and Pleuki. In addition, as I have also previously mentioned, the VC did not take American prisoners, especially in MRII.

We had just leveled off at our cruising altitude when we lost the rear engine. Even though the O-2A has two engines in the push-pull configuration, if you lost the rear engine (push), the single engine capability of the front engine was very marginal. As previously noted, with only the front engine one could only maintain approximately 1100 feet of altitude for sustained level flight. This meant that we were coming down to at least 1100 feet.

We were at night above a solid cloud deck and above mountains in a region where they did not take American prisoners. This did not leave many options for us to consider. Going back to Pleuki was out of the question due to the height of the mountains. If we bailed out and were able to survive the bailout at night, it was very well known that the VC in this area did not take prisoners. As previously noted, the last pilot to go down in this area was found with his Geneva Convention card nailed to his forehead.

I explained the situation to this new young Lieutenant (LT) FAC, since I felt he definitely had a say as to just what course of action we should pursue. I also told him I felt that we did have enough altitude that we could lose and make it to the coast and Nha Trang. I hoped that our slow descent from 8500 feet to 1100 feet would end up over the water east of Nha Trang. If not, the cloud deck would be full of granite.

Being new, the young LT said he would let me make the decision based on my experience. I decided that we would try to make the Nha Trang. We turned southeast and headed toward Nha Trang as we started our slow descent.

God was definitely watching over us that night because after entering the cloud deck, we broke out over the Bay of Nha Trang about two miles east of the end of the runway. Nha Trang was also surrounded by mountains with the base being located in a valley. There are mountains on three sides, and the Bay of Nha Trang is on the east. We leveled off at 1100 feet under the clouds, turned around and landed uneventful at Nha Trang. It took two days to get another engine from Tan Son Nhut to fix the aircraft after which we returned to Saigon.

This was indeed the most hair-raising mission of my tour since it took a while to determine how it was going to end. We had a lot of time to think about all of the possible negative results as we slowly descended down to the cloud deck. There was a tremendous amount of relief when we broke out over water under the clouds.

During my last month in-country (May 1972), we experienced the loss of three FACs. I took two of these losses very personally. The first loss was a young Lieutenant (LT) that I had just checked out. He was flying in Cambodia when he was shot down. I had just completed his checkout the week before. Being the Operations Officer, I had to clean out his room and prepare his belongings to be sent home. It is not a pleasant task to clean out a person's belongings to be sent home.

The next of kin that this LT had listed was his grandmother. In addition to cleaning out his room, I had to write the letter to his grandmother for the Squadron Commander's signature. The letter was to accompany his personal items.

I took his loss very personally since I had just finished his checkout. I wondered if there was something that I had missed during his checkout that I should have covered that would have saved his life.

The question was answered approximately two weeks later. The investigation into this loss determined that he was at approximately 700 feet firing his AR15 out the window during a "troops-in-contact" (TIC) situation when he was hit with three rounds of AK47 in the chest. I knew then that I was not responsible for his loss. I had constantly stressed on every mission that there was no target worth going below 1500 feet and certainly not below a 1000 feet at any time intentionally. This young LT evidently had ignored my words and felt he was bulletproof and that he could win the war on his own. The problem is that two people on the ground with AK47s have you outgunned. They also are stationary while the FAC is flying while firing his weapon out the window. My personal guilt turned more to anger as I accepted his loss as one of being stupid rather than one of omission of my instruction.

The second loss was of a Captain from Laughlin AFB where he had also been an IP. Being from Laughlin AFB myself, I knew him pretty well. I also knew that he was planning on getting out of the Air Force upon return to the States and that he was going to go to law school. In fact, he had already been accepted for law school when he arrived in Vietnam.

Some of the FACs coming to Vietnam had been sent to language school at Monterey, California, to speak French. These FACs were to be utilized in Cambodia where French was more often spoken than English. This Captain was one of those who had completed French Language School.

After I completed checking him out in-country, I decided that I needed to send him to the FOL at Pleiku in MRII because I was lacking experience at this FOL. I was receiving feedback from individuals at Pleiku that some of the young Lieutenants were doing things that they were not supposed to be doing. I told him I needed him to go to Pleiku and rein in these young Lieutenants that were being cowboys. He literally begged me not to send him to MRII. He emphasized that he had been to language school and, therefore, he should fly in Cambodia. Cambodia was less of a risk for FACs as compared to MRII. I told him I really needed him at Pleiku to provide discipline to the young Lieutenants at that FOL. He reluctantly agreed and went to Pleiku.

It was about a week before I left Vietnam when he and another FAC were reported killed in action. I really took his loss personally since I had insisted that he go to Pleiku. I also knew he had a family and of his plan to attend law school upon returning to the States. His loss really bothered me until two months later, when I was in graduate school at the University of Denver, and I found out that he and the other pilot had been doing a low pass and an aileron roll over a FB on his "finni" flight and had crashed. Instead of him providing the discipline to the Lieutenants, the young Lieutenants had won him over to the cowboy way. It was then that my feelings once again turned to anger, and I no longer felt responsible for his loss.

I flew my last mission two days before I left Vietnam. It was on this particular mission, which was in MRIII, when I again lost the rear engine. In MRIII, however, the topography is flat and maintaining 1100 feet did not represent a significant problem while returning to base. I returned to Tan Son Nhut and decided that was enough for me. This was the third engine that I had lost since being in-country. In addition, it was late May 1972, and the North's offensive had pretty much been halted.

In MRI, the South Vietnamese were in the process of pushing the North back into Laos and were regaining control of Quang Tri Province. In MRII, the North had been stopped from capturing a seaport and were retreating back across the border into Laos and Cambodia. In MRIII, the North had been stopped approximately 50 miles north of Saigon and were retreating. During this offensive in 1972, the South Vietnamese forces had successfully defended their country with U.S. assistance provided by U.S. Air Power and the few remaining troops/advisors along with U.S. financial support. (See Figure II for Base locations and other key points of interest.)

I left Vietnam on May 26, 1972, for the States. My tour had been cut short by about a month and a half due to my follow-on assignment. I had been selected for the Air Force Institute of Technology (AFIT) to attend the University of Denver in Denver, Colorado, to obtain a Master's Degree (MS) in Engineering. With the drawdown of the war, the Air Force did not need as many pilots to fly aircraft.

Since my undergraduate degree was Aerospace Engineering, I became a prime candidate to get out of the cockpit and get a MS degree in Engineering. I was scheduled to start classes in Denver in early June 1972. It was going to be tight to get back to Del Rio, Texas, and move my family to Denver and start classes in early June 1972. I was just happy to be going home and very satisfied with my efforts in Vietnam.

An interesting side note occurred when I got the AFIT assignment to the University of Denver. The assignment came down during the time period that I was working all day and flying all night. I was notified that I needed to take the SAT (Scholastic Aptitude Test) test for admission purposes. Here I was in Vietnam, in a very "hot" war, and going back to school definitely was not my primary focus. I also wondered where I could even take such a test in Vietnam.

I went over to Personnel and asked them if they could help me. They said that they thought that the Military Assistance Command, Vietnam (MACV), could help me out. I went over to their headquarters to see if they could provide me with the assistance that I needed. They indicated that they could not provide me with the SAT test, but they could give me the ATGSB test (Advanced Test for Graduate School for Business). I told Personnel that I could only get the ATGSB test and not the SAT. They said take the ATGSB. I scheduled the test as soon as I could get it.

It was scheduled during the time period that I was flying nights in support of An Loc. I really do not remember much about the test, but at least I did complete it. My mind was elsewhere during that time period, and I was not sure that I was even going to make it home. I guess I passed, but I have no idea how I did on the test.

I departed Vietnam after completing 317 combat missions and 630 hours of combat time during the 334 days that I was there. I was looking forward to getting home, but I did not have a clue about the change in attitude of the American Society that greeted me upon my return.

I soon found out just how much the American culture and attitude had changed as far as the Vietnam War was concerned. It was also ironic that my follow-on assignment to graduate school actually set up the circumstance for my return to Vietnam in April 1975.

CHAPTER NINE

The Beginning of the Long Return

Upon my return to the States in late May 1972, I encountered my first impression of just how significantly the American society environment had changed from when I had left. My first impression occurred upon my arrival at the San Antonio Airport. I was wearing my uniform when I exited the aircraft around 2200 hours (10:00 P.M.). My wife and oldest son, who had stayed in Del Rio, Texas, during my time in Vietnam, had driven over to San Antonio to pick me up. As I was entering the airport area, I encountered the first anti-war sentiment toward those of us who had served in Vietnam. A young long-haired individual who was walking by me toward the gate area decided to spit on the floor in front of me as I was walking toward the terminal area. He did not spit directly on me, but the glaring look of disdain that he gave me as he spit on the floor in front of me told me that a significant change in the American general public's attitude had occurred since I had left. I did not say anything of the incident to my wife and son since I was just happy to have survived Vietnam and be home. This incident did however, leave me somewhat confused as to just why this individual would have done such a thing.

My next impression of the significant amount of change that had occurred came as I watched the nightly national news reports on the major networks, especially Walter Cronkite, and their reporting on the war. It seemed that their emphasis was on reporting and showing all the protest that were occuring. It was totally obvious that the biased left-wing liberal news media was totally against the war. This was evidenced by the general anti-war sentiment that they were expressing. I soon discovered that this general anti-war sentiment was not only against the war but also against those of us who had done our duty to serve our country. My culture shock upon my physical return was steadily growing as I tried to process the change that I was experiencing.

My follow-on assignment from Vietnam was to attend the University of Denver in Denver, Colorado. The Air Force had assigned me there to get a MS degree in Engineering. I reported for classes in early June 1972. The University of Denver was and still is a very liberal University.

My attendance in classes and interaction with the other students continued to reinforce my perception of the negative anti-war attitude of the general public toward the military and its members. This confirmed to me that indeed the general attitude of the United States had significantly changed. This negative attitude toward the military fostered a strong feeling of shame of my actions and accomplishments in Vietnam and even my status as an Air Force officer and pilot.

As a FAC in Vietnam, I had been awarded a number of medals. These medals included a Distinguish Flying Cross (DFC), 10 Air Medals, and the Vietnamese Cross of Gallantry with Palm. I received my medals by having them handed to me by another student in a class that he and I were taking. I did not know him, but he knew my name, and he happened to be in the Administration Office when my medals arrived at the University. He volunteered to give them to me since he would see me in class. This method of receiving my medals conveyed to me the idea that these medals did not really mean much and really did not matter. My efforts in Vietnam were really not important. I also watched John Kerry throw his medals over the White House fence. And then I saw Jane

Fonda sitting on an anti-aircraft gun in Hanoi where our POWs were. This was the last straw.

Based on the current environment that I was experiencing, I took my medals home and placed them in the bottom drawer of a steel filing cabinet that was located in a closet in our bedroom. The negative environment that I was experiencing led me to withdraw from any discussion of Vietnam with anyone including my family. I essentially shut down on the previous year of my life.

Going back to college in Engineering in 1972 caused me to have great concern about whether I would be able to compete with the younger generation of students. After all, I had graduated from Iowa State University in December 1964 and had immediately entered the Air Force and pilot training. It was now June 1972, and I had not really even thought about college and the rigors of engineering. I also did not have any experience in the workforce utilizing engineering applications. I was concerned! This concern was erased in one of the first classes that I took. I experienced another incident that again confirmed just how much the general attitude of the younger generation had changed.

It was in one of my first classes on the day before the mid-term exam was to be given when a longhaired hippie type individual tapped me on the shoulder. This individual would come to class barefooted, and he brought his dog with him. The dog (a German Shepherd) would lie at the back of the room and chew on a beer can. I always thought this was strange, but the instructor never said a word to, nor about, this individual. After he tapped me on the shoulder, I turned toward him, and he asked me, "Hey Dude, what does the textbook for this class look like?" I knew right then and there that I could compete with this younger generation.

This type of attitude was totally foreign to me, and I just could not understand it. This attitude was so much different from the one I had experienced at Iowa State University (ISU) in the early 1960s. My undergraduate degree was in Aerospace Engineering, and none of my ISU classmates had an attitude like the one I had just experienced. It did reinforce my feeling that I did not fit well into this different environment, but it did cause me to focus on school and not worry about others.

During the remainder of 1972, while at the University of Denver, I watched the anti-war, anti-military sentiment continue to grow. The United States Government was doing everything it could to withdraw from Vietnam. It also seemed that this anti-war sentiment was the main focus of the news media.

Then on January 27, 1973, our government signed the Vietnam Peace Accord with North Vietnam. A U.S. ceasefire went into effect on January 28, 1973, but we still had military advisors in South Vietnam assisting the South Vietnamese forces in defending their country.

Even with this peace accord, the anti-war sentiment did not stop but continued by demanding that all of our troops come home. Finally, in February 1975, the anti-war supporters and demonstrators were able to apply enough pressure—along with enough anti-war politicians having been elected, that our government voted to cut off all financial aid to South Vietnam. This was when the Vietnam War was lost.

During the Spring Offensive in 1972, the South Vietnamese forces had been able to repel the North with our financial aid and military support. Without our aid, it was only a matter of time before the North would attempt another takeover of the South. By cutting all financial aid, the U.S. indicated to North Vietnam that it no longer cared about South Vietnam and they were free to do as they pleased without interference from us. This action by our government in February 1975, which cutoff all financial aid was like a slap in the face to the military. It seemed to say that all of the actions and sacrifices of military lives just did not matter.

This just reinforced the perception to those of us who served in Vietnam that we should be ashamed of our actions, and we should not talk about it. It reinforced our actions of shutting down and internalizing the conflicting feelings that we had. It was hard for me to understand that we had lost over 58,000 U.S. soldiers during the conflict and now none of that seemed to matter. I had lost a number of friends, pilot training classmates and students whom I had trained as pilots–and for what?

I graduated from the University of Denver in December 1973 with a MS Degree in Civil Engineering. The Air Force assigned me to Kelly AFB in San Antonio, Texas, to work as a Structural Engineer on the C-5A fleet of aircraft. I soon worked my way up to be the Lead Structural Engineer for the C-5A fleet for the Air Force. Kelly AFB was the System Program Manager for the C-5A fleet. My position as the Lead Structural Engineer is what resulted in my return to Vietnam in 1975.

Since our government had cut off all financial aid to South Vietnam in February 1975, it was not long before the North launched another major offensive to take over the South. Without our aid and military advisors, it became obvious that the South would not be able to successfully defend itself as they had in 1972. The U.S. began a program of evacuating U.S. civilians and embassy personnel from Vietnam in April 1975.

It was on one of these evacuation missions from Vietnam when a C-5A crashed. This mission had been named "Operation Babylift." In addition to a number of civilian embassy employees who were being evacuated, the U.S. had agreed to evacuate a number of orphaned children. Many of these children had been fathered by U.S. soldiers and then abandoned by their mothers after our military forces withdrew. This C-5A crashed on April 4, 1975. Being the C-5A Lead Structural Engineer, I found myself headed back to Vietnam to investigate this accident.

CHAPTER TEN

"Operation Babylift"

THE 4ᵀᴴ OF APRIL 1975 started for me at 0500 hours (5:00 A.M.) with a phone call from the Kelly AFB Command Post. When anything significant occurred with a C-5A aircraft, as the Lead Structural Engineer for the C-5A fleet, I was on the list to be notified. The Command Post notified me of the C-5A accident in Vietnam. I arrived at my office at 0700 hours (7:00 A.M.) that day anticipating that I would be required to complete some activity as a result of this accident. I did not anticipate that I would be assigned to the Accident Board to investigate the accident.

At approximately 0900 hours (9:00 A.M.), the Command Post again called me to inform me that Major General Kelly (The Kelly AFB Air Logistic Center (ALC) Commander) had selected Major Russ Gregory (a Mechanical Engineer for the C-5A fleet) and me to be on the Accident Board. The Accident Board President was to be a 2-Star Major General from the Military Airlift Command (MAC). Having a 2-Star General selected to be the Board President indicated just how important the investigation of this accident was to the Air Force. I do not know of any other Air Force accident investigation teams being headed up by a 2-Star General. There may have been, but I did not know of any.

My selection to this Accident Board turned out to be the second major potential crossroad in my Air Force career. I do not know if it was just coincidence or not, but both of these crossroads had to do with personnel from the Military Airlift Command (MAC).

The first major crossroad was in Vietnam with the Colonel Wing Operations Officer and the Airlift Wing at Tan Son Nhut. The second major crossroad was with this 2-Star General who also was assigned to MAC as the Deputy Commander for Maintenance for MAC.

The Commander of MAC, a 4-Star General, requested two engineers from Kelly AFB be provided to be part of the accident investigation team. This was due to the fact that Kelly AFB was the Program Manager for the C-5A fleet. The Command Post also informed me that a C-135 aircraft from MAC Headquarters at Scott AFB, Illinois, would arrive at 1200 hours (12:00 P.M.) that day to pick up Major Gregory and me to go to Vietnam. I immediately went home to pack a bag in order to get back to the base for the 1200-hour departure.

As luck would have it, my mother-in-law and her sister had just arrived from Iowa to visit my wife and me. I rapidly packed my TDY bag and told my wife I was heading back to Vietnam, and that I did not know how long I would be gone or when I would be back. My wife was used to this, but my mother-in-law and her sister were not. My wife just said, "Okay," but the looks on my mother-in-law and her sister were priceless. My wife told me later that after I left, they both wanted to know how often something like this happened. My wife simply replied that she was used to it and went about her business. My wife is a very remarkable woman.

We departed Kelly AFB at 1200 hours (12:00 P.M.) on April 4, 1975, heading for Clark AFB in the Philippines. The Accident Board consisted of approximately 35 board members headed by the 2-Star Major General Board President. A full Colonel from MAC had been chosen as the Investigation Officer. We had to stop at Clark AFB in the Philippines, since when we arrived in Vietnam, we would be counted as part of the U.S. troop strength level allowed to be in-country by the Paris Peace Accord Agreement that was now in effect.

Because of this Peace Accord, only seven of us from the Accident Board team would be allowed in-country at any one time in order to stay below the strength level allowed in-country by the Accord. This was just the beginning of the difficulty in conducting this accident investigation.

The next day, the seven of us selected by the General to go, were flown to Tan Son Nhut AB in Vietnam on a C-130 aircraft to begin the investigation. We also were not allowed to remain in-country overnight, so we had to be flown back to the Philippines each night by the C-130. In addition to not being able to stay overnight, and the fact that only seven of us could be in-country, we were not allowed to wear any rank on our fatigue uniforms and we could not carry any weapons.

After arrival at Tan Son Nhut, Air America helicopters of the Central Intelligence Agency (CIA), transported us to the crash site. We essentially arrived on site 24 hours after the accident had occurred. Parts of the aircraft were still burning due to the jet fuel that was still present. The CIA and South Vietnamese personnel were still in the process of recovering remains. The survivors had been rescued the day before just after the accident.

The aircraft had broken into four sections on impact. These sections were the tail section, the troop compartment section, the wing section and the cockpit section. The cargo section of the fuselage no longer existed. The entire cargo section of the aircraft had been totally destroyed on impact. Only the troop compartment, which is located above the cargo area, 19 feet above the cargo deck floor, had survived the accident.

This troop compartment, which had 80 airline-style seats, was now sitting on the rice paddy in its final resting place. The cockpit section and this troop compartment section is where the survivors of this accident had been located. Some of the survivors included the pilot, copilot, and loadmaster. Without these survivors, we may never have been able to determine the cause of this accident, especially in the time period that was available and location of the accident.

We also did not know the actual number of souls that were on board. All we knew was that there were greater than 300 people including a number of babies. We were later informed of the actual number of babies and older children who were on board. We had 145 babies and 102 older children—some with physical handicaps.

The 145 babies had been placed in the troop compartment seats with as many as two or three babies to a seat. The 102 older children and at least 50 U.S. Embassy personnel along with other civilians had been seated on the cargo deck floor. These passengers had been strapped down with cargo ties since there were no seats in the cargo area. The entire cargo section where all of these people were located, no longer existed. All of these lives were lost when the fuselage cargo section was totally destroyed upon impact. What we did know is that 175 people did survive, but more than 150 people had been lost.

Upon our arrival in Vietnam, it was quite obvious that South Vietnam was going to be taken over by the North. The C-5A accident site was approximately 20 miles northeast of Tan Son Nhut AB. While at the crash site, we could hear the war going on and see smoke rising from many of the villages that were located relatively close to Saigon. In addition, the physical conditions at Tan Son Nhut had greatly deteriorated from when I departed in May 1972. The area that I had worked out of was almost non-existent due to the amount of deterioration that it had experienced. It definitely did not look good for the future of South Vietnam.

In addition to the ongoing war, the small number of Accident Board members allowed in-country at any one time, and the fact that we had to commute between the Philippines and Vietnam each day, we had other significant obstacles to contend with while conducting this investigation.

The first was that the accident site itself could not be secured. Since we were not allowed to carry weapons, the South Vietnamese said they would address our security needs. The troops that we were assigned to provide security turned out to be 12-14 years old barefooted boys in tattered clothing with AK-47s. So much for security during the day, and there was no security for the site at night. This was also a good indication of the deteriorating quality of the South Vietnamese forces.

On the first day of our arrival at the crash site, the four sections of the aircraft were, for the most part, intact. The cockpit section and the troop compartment had not been disturbed. By the second day, however, all of the seats in the troop compartment had disappeared along with some of the cockpit instruments. By the third day, all of the headliners in the troop compartment and all of the oxygen masks that had deployed when the rapid decompression had occurred, were gone. By the fourth day most of the aircraft wiring and aluminum tubing had been taken, and exterior metal skin sections of the aircraft structure were now disappearing.

Without proper security, the local population was salvaging aircraft parts and metal to sell to junkyards or to keep for their own use. As many as 150-200 locals would be present each day salvaging parts as we tried to investigate the accident. The aircraft wreckage was disappearing right before our eyes, and the locals salvaging parts greatly outnumbered us. As the Lead Structural Engineer from Kelly AFB, I was responsible for disposal of the wreckage. Because of the locals and the war, this turned out not to be a problem.

Since we did not have a complete manifest of the individuals who had been present on the cargo deck, the State Department was placing a significant amount of pressure on us to identify the human remains that were present under the troop compartment of those individuals who had been seated on the cargo deck floor.

As the structural engineer, I was given the responsibility by the 2-Star General to get to these remains in order to accomplish this tasking. The most obvious method was to try to flip the troop compartment over. To attempt this action, the CIA provided two helicopters with cables that could be attached to hooks on the bottom of their helicopters. Due to the weight of the troop compartment, however, we were not able to accomplish this without tearing the hook supports out of the bottom of the helicopters. Another course of action was required.

The next course of action was to cut out each section of the troop compartment floor, which was also the ceiling of the cargo area. With the removal of each 2'x4' section of flooring, we finally gained access to the remains that were underneath. One can imagine what we encountered once we gained access to this area.

We did, however, find a couple of nametags, some purses and billfolds, but not much else that would allow identification of remains. After two days of this activity, with minimal results, the State Department agreed that this effort would not produce the desired results. It took me approximately two months to get the stench out of my mind from this effort. Something like this does leave a lasting impact on one's mind in more ways than one.

After approximately a week of commuting between the Philippines and Vietnam, we finally got clearance for all 35 Accident Board members to be in-country and remain overnight. This helped immensely in our efforts to investigate the accident.

With the complete Board present, we were able to conduct interviews of the survivors such as the pilot, co-pilot and loadmaster. The information that we gained was significant. We learned that at approximately 23,000 feet on climb-out during departure over the South China Sea, an explosive rapid decompression occurred. This explosive rapid decompression was caused by a catastrophic failure of a section of the aft ramp and pressure door, which departed the aircraft. When the ramp section and pressure door were blown out of the aircraft, the majority of the flight control cables were torn out along with the rupture of the hydraulic lines running to the rudder and elevator controls of the aircraft. The pilot was left with one aileron and the four engines, which had not been impacted, to control the aircraft. It was only with outstanding airmanship and the grace of God that there were any survivors from this accident.

This information allowed us to focus our attention on the aft ramp area and the aft ramp locking system. We began searching for any parts from the remaining aft ramp section and any parts from the locking system that made it to the crash site.

The C-5A aft ramp locking system consists of seven locks on each side of the aft ramp. Each lock consists of a stirrup or clevis, which is attached to the ramp (See Figure 3). These seven stirrups are U shaped and have a one-inch steel pin in the open end that is engaged by a bell crank hook which is located on the fuselage. These seven bell crank hooks are tied together by six tie rods that connect to a single hydraulic actuator (See Figure 4). When the actuator is activated, it moves all seven bell cranks forward to engage the seven stirrups on the ramp to affect the locking activity. The six tie rods between the seven bell cranks have to be of a precise length to ensure all seven locks obtain an over center and a locked condition. The tie rod lengths are critical for this locking activity to be completed.

We now knew we needed to find the 14 aft ramp locks which consisted of 14 bell cranks (hooks), 14 stirrups, 12 tie rods and two hydraulic actuators. We did, however, have a problem. Part of the aft ramp and pressure door were at the bottom of the South China Sea.

The Board President contacted the Navy, and being a 2-Star General, he did not have much trouble getting the Navy's assistance in this matter. We were able to provide the Navy with a relatively precise location as to where we thought the aft ramp section and pressure door should be located. The Navy went to work searching the floor of the South China Sea.

Meanwhile, the majority of the Accident Board members continued to search for the parts that we most desperately needed. We also were looking for the Crash Data Indicator Positioning Recorder (CDIPR), which is the black box for the C-5A aircraft.

The CDIPR is located on top of the tail section and will deploy if it experiences three G's or is submerged in water. This black box monitors all of the aircraft flight and engine parameters and has the last 30 minutes of cockpit voice recording. The CDIPR was not at the crash site, and we could only assume that one of the locals had found it and took it home to sell or use. It did look like a tape recorder. As I mentioned previously, the locals greatly outnumbered the Accident Board in retrieving parts from the crash site.

It was during our second week in Saigon, and the war was not going well for the South, when two South Vietnamese F-5E pilots out of Bien Hoa attempted to bomb the South Vietnamese Presidential Palace. Soon after that, the South Vietnamese President resigned. Things were really now not looking good for South Vietnam.

During our investigation, those of us at the accident site would occasionally see NVA soldiers come out of the bordering tree lines and observe us. In addition, they would occasionally fire a round over our heads just to let us know that they were there. I did not feel that they were going to attack us since, if they did, I think they thought that this might bring the United State Military back into the conflict. It is sad to say, however, if we had been attacked, I am now not so sure that this would have been the U.S. response based on the amount of anti-war and anti-military sentiment that was present in the United States.

I do, however, remember one particular day that I think may have been the closest that we came to being attacked. We were at the crash site, and it was during the time period that we could only have seven members present in-country. There were seven of us along with the normal 150-200 locals gathering parts. Around 1100 hours (11:00 A.M.) the locals started disappearing from the crash site as did our security guards. As we looked around, there were only us seven Board members left. Having been a FAC, I had been given the responsibility of handling our radio. We had a radio for contact with the CIA Air America pilots if we needed it.

With us seven being the only ones left at the crash site, it was not difficult for me to convince the 2-Star General that something was not right. He agreed, and I contacted the Air America helicopter pilots and requested an immediate extraction from the crash site. Within five to seven minutes, two helicopters were there and we departed the crash site uneventfully.

The next day when we returned to the crash site, the North Vietnamese had broken the rice paddy dikes and reflooded the rice paddy where we were working. We had closed the dikes to drain as much as water as possible from the rice paddy to assist in locating parts. With the reflooding of the rice paddy, the water hampered our search for parts. In many places the water was thigh deep. This caused us to have to use metal polls to tap the ground until we contacted a metal object. The locals were doing the same thing.

An interesting incident occurred one day while we were contending with the water problem. One of the local young boys, who was also searching for metal parts evidently, contacted an unexploded ordinance that went off. The blast blew off his foot. An adult man carried this boy over to me and indicated he needed our help. Since I had the radio, I called Air America and requested a medevac extraction for this boy. An Air America helicopter arrived and medevacked the boy back to Tan Son Nhut.

I do not know what happened to this boy, whether he survived or not, but I thought it was interesting that the locals brought the boy to the Americans for help. After the man handed me the boy, he turned around and walked off. He seemed to know that we would take care of the boy. The image of this boy still remains clearly in my mind 40 years later.

Since the locals had removed a significant number of parts, and we had not found the CDIPR and we still needed key parts from the aft ramp, we initiated a program as an attempt to recover these key items. We began searching junkyards and other salvage facilities where the locals may have taken the salvaged parts to sell. We also printed "pointee-talkie" sheets with pictures of the parts that we were looking for along with how much we would pay for them. We began going around to these junkyards showing the "pointee-talkie" sheets. This effort did provide a good result.

A French reporter contacted us and indicated that he could get the CDIPR for us. We had indicated that we would pay $200,000 piasters, which was about $200.00 in American money. The French reporter did provide us with the CDIPR but it was not in good condition. It was obvious that someone had tried to get the tape recorder to work, but did not succeed. The cockpit recording tape was there but not on the take-up reel. We were, however, able to send the tape and the black box to the Federal Bureau of Investigation (FBI) Laboratory in Washington, D.C., for analysis. They were able to retrieve most of the information that we had hoped for from the black box.

We investigated the C-5A accident as best we could under the deteriorating war conditions that were occurring. Without the U.S.'s financial aid that had been cutoff in February, the South Vietnamese government was collapsing around us. Congress also did not approve the 720 million dollars of emergency aid that President Ford had requested when this offensive began. The U.S. had turned its back on South Vietnam.

We were very lucky in that two days before we left Vietnam, the Navy found the aft ramp section and pressure door at the bottom of the South China Sea. Once we got these sections, we had enough of the pieces of the puzzle to determine the cause of this accident.

Since it was obvious that South Vietnam was going to fall to the North, the Accident Investigation Team packed up the recovered aircraft parts that we had at that point and left. We departed on the last two fixed-wing aircraft to leave Tan Son Nhut Air Base. These two aircraft were C-141s that the 2-Star General had ordered to transport the Team and the recovered parts out of Vietnam. The next day the North Vietnamese cratered the runway.

We left Vietnam on April 27, 1975, and South Vietnam fell to the North on April 30, 1975, three days after our departure. Needless to say, even though I was responsible for disposal of the wreckage, the locals essentially had taken care of that. What was left was left to the North Vietnamese. The only wreckage retained were the two C-141 aircraft loads that we left with from Vietnam.

The Accident Investigation Team left Vietnam and landed at Clark AB in the Philippines. We felt that we had enough information to start analyzing the parts that we had. The Navy's recovery of the aft ramp section and pressure door from the bottom of the South China Sea provided us with a tremendous amount of key information as to the cause of this accident.

The section of the aft ramp that the Navy recovered had three of the aft ramp stirrup (clevises) still attached. Two of these stirrups were in perfect condition while the third had only minor damage. These were the first three locks on the right side of the aircraft's aft ramp. This pointed to the idea that at least two of the locks had unlocked, and the third was also in the process of unlocking when the catastrophic failure of the aft ramp occurred. We had found the other four right side aft ramp locks at the crash site. They were all significantly damaged.

The aft ramp locking system on the C-5A is designed so that the aft ramp and pressure door can withstand the loads from the internal aircraft pressurization forces with three of the locks unlocked as long as these three locks are not consecutive. If three every other locks are unlocked, the remaining four locks can maintain the pressurization loads experienced on the ramp generated by the aircraft pressurization system. In this case, it appeared that the first three locks on the right side of the aircraft had unlocked resulting in catastrophic failure of the aft ramp. The unlocking of these three locks must have occurred as the aircraft was climbing through 23,000 feet as the cabin pressurization forces were building.

When these three locks unlocked, the pressurization forces being held by these three locks were immediately transferred (dumped) to the remaining four locks. With this condition, the surface of the aft ramp was not able to withstand all of the forces from the four locks; and catastrophic overload failure of the aft ramp occurred between the third and fourth lock. The aft ramp split apart due to the pressurization forces, and the right side of the ramp was blown downward as it pivoted around the left side that remained locked.

Since the pressure door is connected to the aft ramp, the right side of the pressure door also rotated downwards. The pressure door has four large steel finger supports that rest on four large stainless steel rollers that are located on the fuselage at the top and end of the cargo area. This area of the fuselage above the rollers contains the flight control cables and hydraulic lines that go to the tail section of the aircraft.

As the aft ramp split and pivoted around the left side locks, it pulled the pressure door down with this pivot action which caused the steel pressure door fingers to be pulled down off the steel rollers from right to left as the ramp and pressure door pivoted down and away from the aircraft.

The steel fingers of the pressure door are what caused the damage to the aft fuselage area where the flight control cables and hydraulic lines are located going to the tail section. This all occurred in a second or less when the three locks unlocked and dumped their loads onto the remaining four locks. This is why an explosive rapid decompression occurred in the aircraft.

We now understood why the explosive decompression had occurred and the failure pattern of the aft ramp and pressure door. We also knew why the flight controls and hydraulic lines had been disabled. This failure pattern did, however raise a number of other questions.

One question was, why did this occur on the third takeoff after leaving Travis AFB where it had started this mission? The aircraft had landed once at another location before proceeding to Vietnam. The aircraft had picked up a load of 105 Howitzers to take to Vietnam. This failure did not occur on the second takeoff, so why did it occur on the third takeoff? These were questions we needed answers to, and we needed to go to Travis AFB to get the answers.

We had ruled out sabotage in Vietnam with bomb-sniffing dogs. We needed to rule out the possibility that a bomb may have been placed in one of the passenger personal bags which had been loaded just prior to takeoff from Tan Son Nhut. Many of these bags had been placed on the aft ramp for the trip.

After a week at Clark AFB in the Philippines, the Team was now on their way to Travis AFB, California. This is where the aircraft had been assigned and where it started this fateful mission. We needed a lot more information in order to answer our questions. We needed to review the aircraft's 781 Form, which is the aircraft's maintenance records, and to talk to the maintenance personnel. We were beginning to piece together just what had caused the failure sequence, and we needed to know why the failure had not occurred on the earlier takeoff from Travis AFB or the takeoff from the intermediate stop.

Review of the aircraft's 781 Form revealed that this particular aircraft had been in what is called a "Cann Bird" status prior to its initial takeoff from Travis AFB. What this means is that this aircraft was being utilized as a spare parts store for maintenance. When supply did not have a spare part available, the part was cannibalized from this aircraft to be used on another aircraft.

During the C-5A procurement process, the program was having both cost and weight-overrun problems. In an attempt to stay within the budget, cuts were made. Many of the cuts made were by not buying all of the spare parts needed to maintain the aircraft. The idea was to buy the spare parts at a later date. To off-set the lack of spare parts, MAC had instituted a program whereby one aircraft would be placed in a status of non-flying in order for parts to be cannibalized for a time period, then replacing all of the cannibalized parts in time to fly the aircraft as required. MAC was required to fly its assigned aircraft at least once every 30 days. This was the aircraft used as the "Cann Bird" for 20 days, and then during the last 10 days, maintenance personnel would replace all of the cannibalized parts to get the bird ready to fly. The trip to Vietnam was this bird's first mission after coming out of "Cann-Bird" status.

Review of the 781 Form revealed that 26 cannibalizations had been completed on this bird prior to putting the parts back on the aircraft. Two of these cannibalizations involved the removal of the two tie rods located between locks 2 and 3 and 3 and 4 of the right side aft ramp locking system. This was a very big red flag. We now knew we had to interview the maintenance personnel who had replaced these two tie rods.

It is important to understand just how the C-5A aft ramp locking system was designed to work. As I have noted, the system is a gang locking system of seven locks with one hydraulic actuator.

When the actuator is activated, it drives the bell cranks (hooks) forward to engage the stirrups to a position that is 3° to 7° over center as the locked position. All seven locks have to reach the 3° to 7° over center position to be considered locked.

To do this, the six tie rods between the seven locks have to be of a precise and specific length between the bell cranks to ensure that the proper over center position is achieved for each lock. To ensure that these tie rod lengths are of the proper length, the aft ramp locking system has to be properly rigged to ensure that all seven locks reach the desired over center and locked position. If any single locking component in this system is removed, all seven locks and six tie rods have to be re-rigged to ensure that the desired over center and locked position of 3° to 7° is obtained for all locks.

This re-rigging maintenance procedure is a long and tedious procedure to say the least. In fact, the re-rigging procedure for this maintenance activity requires 35 pages of instructions in the C-5A Aircraft Maintenance Manual—the T.O.–C5A–2-12 Manual. It also requires approximately eight hours to complete the re-rigging procedure for each side of the ramp.

Discussion with the Maintenance Personnel who had performed the replacement activity to these two tie rods revealed the following. One Maintenance Mechanic had started the tie rod replacement procedure approximately half way through the maintenance work shift. A second Mechanic on the next shift finished the job and had assumed that all the re-rigging instructions had been completed on the first shift. The first mechanic, however, had merely replaced the tie rods but had not completed the re-rigging procedure to verify that the tie rods were of the proper length between the locks.

The way the system works is that if the locks are between the 3° to 7° over center position, when the aircraft pressurization forces begin to build on climb-out, the forces being developed in the tie rods actually tries to force the locks into a greater over center position. If the locks are not over center, in the 3° to 7° position, when the aircraft pressurization forces begin to build on climb-out, the forces being developed in the tie rods actually try to drive the locks to an unlocked position. This is why the re-rigging procedure is so critical when the ramp locking system is disturbed.

Some may ask, "Is a gang locking system the best way to design a system, or is it better to have individual actuators for each lock?" Of course, the answer is, "Yes, a single actuator for each lock is best." When the C-5A was being designed, however; it was having problems with weight during the design phase. By using a system that utilized only one actuator for seven locks, the weight of the six additional actuators on each side of the ramp was saved.

It should be noted, that due to the weight problem during the procurement process of the C-5A aircraft by the Air Force, Lockheed Aircraft Company was given a $300.00 incentive for every pound of weight it could save. The savings of six hydraulic actuators per side provided a large incentive to persuade Lockheed to use the gang locking system for the C-5A. Although not the best design for a locking system, it is an adequate design if properly maintained. It does, however, require a lot of maintenance activity to maintain such a system. This locking system was also the beginning of the conflict between Major Russ Gregory and me and the 2-Star General Accident Board President.

At Travis AFB we found out that the aft ramp had been locked when the bird was placed in "Cann Bird" status. We also learned that since the aft ramp re-rigging procedure had not been completed, the aft ramp had not been opened until arrival in Vietnam where the 105 Howitzers were unloaded.

At the intermediate stop after leaving Travis AFB, the 105 Howitzers had been loaded through the front visor loading ramp system. The aft ramp had not been utilized until Vietnam when it was opened to unload the Howitzers. This meant that the aft ramp had not been utilized since prior to the cannibalization activity of the two tie rods when the bird was in "Cann" status. This explained a lot as to why the failure had not occurred on the first two takeoffs from Travis AFB and the intermediate stop.

The interview with the Load Master after the accident indicated that he did have some problem getting a locked condition at Tan

Son Nhut just prior to takeoff. He stated that it took him three or four times to get a locked green light indication when he closed the aft ramp at Tan Son Nhut. This did cause him some concern, but he was able to finally get a green light locked indication for the ramp.

Each lock has a squat switch that is supposed to indicate when each bell crank has achieved an over center (3° to 7°) and locked position. These squat switches, however, are mounted on the fuselage by thin metal clips. These clips are easily damaged or bent.

While at Travis AFB, the 2-Star General Board President began to hold daily meetings at 0900 hours (9:00 A.M.) each day to discuss the progress of the investigation. It was during these daily morning meetings that Russ Gregory and I got in trouble with the 2-Star General.

Each day the 2-Star General would come into the room and go directly to the blackboard and with a piece of chalk, he would write in large letters, "Design Deficiency." The discussion as to the actual cause of this accident was beginning to manifest itself in two possible scenarios. The scenario that the General was supporting was "Design Deficiency," due to the gang-locking system, which only had the single actuator. He felt that each lock should have had its own actuator.

The second theory about the cause of the accident was developed by Major Gregory and me along with discussion with engineers from Lockheed. We contended that the cause of the accident was due to "Maintenance" as a result of not completing the re-rigging procedures on the right side aft ramp locking system during the replacement of the two tie-rods—not "Design Deficiency." We felt that the design, although not the best, was adequate when considering the weight problem of the C-5A during procurement. Our theory as to the cause of the accident was based on how the locking system actually works.

Since the aft ramp was locked and closed when the aircraft was placed in "Cann Bird" status, all seven locks were in the over center position of 3° to 7°. The ramp was properly rigged prior to the last closure operation. Since the aft ramp had not been operated during the time it was in "Cann" status, the locks remained in the over center and locked position. This is why the ramp did not fail on the takeoff from Travis AFB.

In addition, since the front visor loading system was utilized during the loading of the Howitzers at the intermediate stop, the aft ramp again was not opened, and, therefore, the failure did not occur after the second takeoff. The aft ramp system was, however; utilized at Tan Son Nhut when the Howitzers were off-loaded. The aft ramp was opened to conduct the off-loading activity. This was the first time that the aft ramp system had been utilized since prior to the aircraft being placed in "Cann Bird" status.

As I have previously noted, when the locks are in the 3° to 7° over center position, the forces developed in the tie rod system as the cabin pressurization builds on climb-out after takeoff, actually try to force the locks into a greater over-center and locked position. This is why there was no problem on the first two takeoffs. When the locks or some of the locks, however, are not in the 3° to 7° over center position or actually not over center at all, the forces in the tie rod system during climb-out are such that these forces try to drive the locks toward an unlocked position.

Our theory was that after the ramp was opened in Vietnam and then closed just prior to takeoff, the closure occurred with a non-rigged system. When this happened, the actual position of locks one, two and three was now in doubt as to being in the 3° to 7° over center and locked position. This is why the Load Master said he had to try three or four times to get a green light locked indication at Tan Son Nhut prior to takeoff.

What we theorized was that there was an improper length of the two tie rods between locks 2 and 3 and 3 and 4 after their replacement, and the system was not re-rigged. We theorized that on the closing of the ramp and the attempt to lock the ramp, lock 1 did not achieve an over center position at all. We felt that lock 2 only achieved a position of top dead center and could go either way as the forces built up on climb-out. Lock 3 may have achieved a position of just slightly over center, but it was less than 3° over center.

After takeoff and as the cabin pressurization forces were building, the forces developed in the tie rods between locks 1, 2 and 3 were building in such a manner as to drive these locks toward the unlocked direction. Since lock 1 was already unlocked and lock 2 was top dead center, as the forces were building, they were sufficient to cause lock 3 to move to an unlocked position and also pull lock 2 to an unlocked position. When this occurred, the loads from locks 1, 2, and 3 were dumped onto the remaining four locks. This resulted in exceeding the structural limit of the aft ramp between locks 3 and 4, which resulted in catastrophic structural failure of the aft ramp. This resulted in the previously noted failure pattern and the explosive rapid decompression as the aft ramp and pressure door departed the aircraft.

We supported our theory based on the evidence of the perfect condition of the first two stirrups found on locks 1 and 2 on the ramp section retrieved by the Navy from the South China Sea. This, along with the evidence of minor damage to lock number 3 stirrup, further supported our theory. The 2-Star General did not agree with our theory and was pushing "Design Deficiency" as the cause.

The 2-Star General was a large man at 6'4" or 6'5", and with 2 Stars, he was very intimidating just by his physical appearance. As such, most of the Accident Board members were in tune to supporting the General's position. As engineers, however, Russ and I just could not go along with his position.

Each day after the General wrote "Design Deficiency" on the blackboard, either Russ or I would raise our hand and express the position that we could not support that cause of failure. What followed after our expression of non-support was a verbal tirade from the General until he would end the meeting. Russ and I would take turns each day on who would raise their hand after which the tirade would occur.

This went on for a week with each 0900 hour (9:00 A.M.) meeting going the same way. Then on Friday of that week, after his verbal tirade, the General said that if we could not support him, he no longer wanted Russ and me on his Accident Board. Russ and I said that if we were dismissed from the Accident Board, we would have to write a dissenting report to his accident report. He responded with, "Do whatever you think you have to do." Russ and I departed Travis AFB and returned to Kelly AFB. Since the C-5A wreckage parts were the responsibility of Kelly AFB, we essentially took our parts and went home.

We had now placed ourselves in a situation where two Majors were disagreeing with a 2-Star General. Usually, in cases like that, the two Majors do not fair very well. Our careers were on the line. We could only hope that we could get the support of our 2-Star General, the Air Logistics Center (ALC) Commander, at Kelly AFB. In addition, he would have to get the support of the 4-Star General who was the Commander of the Air Force Logistics Command (AFLC). We knew that the 2-Star General Board President would have the support of the 4-Star General Commander of the Military Airlift Command (MAC).

The background history of the Board President 2-Star General does provide some insight into why he was so determined to reject our theory. I felt he had been placed in a no-win position when he was selected to be the Board President for this accident. When this 2-Star General was a full Colonel, he was the Air Force's C-5A System Program Office (SPO) Director that was responsible for procuring the C-5A aircraft for the Air Force. The C-5A SPO office was part of the Air Force Systems Command (AFSC). As the C-5A SPO Director, he was definitely aware of the weight problem of the C-5A and knew about the weight reduction $300 per pound incentive provided to the Lockheed Aircraft Company. In fact, as the C-5A SPO Director, he probably had to approve this reduction in weight for the single hydraulic actuator locking system.

Once the aircraft fleet was procured, the C-5A fleet was assigned to an operational command. The C-5A fleet was assigned to the Military Airlift Command (MAC) for operations. In addition, once an aircraft system has been assigned to an operational command, the program management of that fleet is transferred from AFSC to the Air Force Logistics Command (AFLC).

The Air Logistics Center (ALC) at Kelly AFB was assigned as the Maintenance Depot and Program Manager for the C-5A fleet. This responsibility for the C-5A fleet then rests with the assigned ALC until the aircraft system is retired for service. Being assigned to Kelly AFB in support of the C-5A fleet is how Major Russ Gregory and I became involved with this accident.

The problem for the 2-Star General resulted from the fact that at the time of the accident, he was the MAC Deputy Commander for Maintenance for the entire command. This meant that he was responsible for all maintenance activities performed by MAC personnel. So the problem for this 2-Star General was to decide if he was not doing his job during the C-5A procurement process or that he was not doing his job currently as the MAC Deputy Commander for Maintenance. I think he decided to support the idea that he was not doing his job as C-5A Director as opposed to not doing his job now. I also felt that as a 2-Star General, he was not used to having people telling him "No."

Russ and I returned to Kelly AFB and, of course, had to report to General Kelly. General Kelly was the 2-Star General in command of the San Antonio ALC. He was also a large individual physically at 6'4"or 6'5" himself. This also made him a very intimidating individual just by his physical appearance. In addition, he had been a WWII Prisoner of War (POW). Russ and I both had a lot of respect for General Kelly.

We briefed General Kelly on the accident investigation and on the two theories as to the accident cause. Fortunately for us, General Kelly felt that our theory was the proper cause of the accident. We also informed him we had been dismissed from the Board by the 2-Star General Board President for non-support. General Kelly said that he would support us submitting a dissenting report to the Accident Board's report. In addition, since we had all of the wreckage parts, General Kelly made a vacant hangar available to us and told us to prove our theory, which then became known as the San Antonio Position.

I am not sure but I always had the feeling that General Kelly did not really care for the 2-Star Board President. He never said anything negative about him but his body language when he spoke of him revealed a dislike of the man.

Russ and I began our efforts to prove our theory of the accident cause. We ran numerous laboratory metallurgical analyses of the failed wreckage parts to determine the type of failure mode for each of the failed surfaces. We also built a mock-up of the aft ramp locking system in order to test our theory of the forces developed in the tie rods as cabin pressurization forces built on climb-out after takeoff. This was necessary in order for us to prove our theory as to position of locks 1, 2, and 3 after ramp closure and why these locks were unlocked or became unlocked in order to start the failure sequence. Our efforts proved successful, and we were able to duplicate the failure sequence to within 300 feet of the actual altitude that it occurred.

With General Kelly's approval, we began writing our dissenting report for submittal to the Air Force Safety Office at the Pentagon for their consideration. The Air Force Safety Office would have the final say as to the cause of the accident.

When the 4-Star General Commander of MAC was informed that San Antonio was indeed going to submit a dissenting report, things became very uncomfortable for Russ and me. General Kelly assured us that he had our backs and not to worry. We now had a situation where two Majors had said "No" to a 2-Star General, and another 2-Star General was telling a 4-Star General that he agreed with the two Majors. The lines had definitely been drawn in the sand.

Before we completed our report to be submitted to the Air Force Safety Office, the 4-Star MAC Commander requested a meeting of all parties involved. This meeting was held at Wright Patterson Air Force Base in Ohio. The situation was now at the 4-Star General level.

This meeting consisted of the following individuals: the three 4-Star General Commanders from MAC, AFSC and AFLC, our 2-Star General and the 2-Star General Board President. In addition, Russ and I, the 2 Majors, were also present.

The five General officers sat at a round table in a small meeting room. Russ and I sat in chairs behind General Kelly. There were 16 stars in that room along with two lowly Majors. This meeting was very intense. Talk about intimidation, this was it.

The 2-Star Board President started the meeting off with his version of the accident cause and provided his justification. General Kelly then expressed the San Antonio Position as the cause in a very calm manner without once looking at the Board President. He informed the 4-Star Generals that our mock-up of the aft ramp locking system had verified the failure pattern and that we had been able to duplicate the failure sequence. This, of course, got to the Accident Board President, who began to raise his voice in rebuttal. The one comment that I distinctly remember him saying was, "I don't care what those two Majors say; the cause was "design deficiency." Russ and I were not asked to say anything. General Kelly did all of the talking for us.

The results of this meeting ended with the MAC Commander supporting his Board President. The AFLC Commander expressed support for General Kelly and the San Antonio Position. The AFSC Commander did not support either side. I think he wanted to wait to see what the Air Force Safety Office had to say. If the Safety Office accepted the Board President's version, the cause would be focused on him and AFSC. If the San Antonio Position was accepted, the cause of the accident would be focused on MAC and not AFSC. He decided to wait and see. This resulted in what could be called a "Hung Jury." We all returned to our respective bases with no resolution as to an agreed cause of this accident. Russ and I finished our report, and General Kelly submitted it to the Air Force Safety Office at the Pentagon. All we could do now was wait to see which way the Air Force Safety Office would rule.

This turned out to be a long six-month wait for their final ruling, especially for me. Russ Gregory had been enlisted before going to AFIT to get his engineering degree and commissioning. He had reached retirement eligibility during this time period and retired from the Air Force. This left me alone awaiting the answer from

Air Force Safety, which would determine if I had a career to finish. I was not eligible for retirement and would not be for a while.

The waiting was very stressful, and I had to wonder if we had chosen the correct path. Majors normally do not do well when saying "No" to Generals. Generals are very used to getting their way on things. With 4-Star Generals involved, I just was not sure how it was going to turn out. Politics play a large roll in relationships among General officers. I just hoped that the Air Force Safety Office would not play politics.

The thickness of the Accident Board's Report consisted of two volumes each three inches thick. The San Antonio dissenting report consisted of one volume that was three inches thick. This was a lot of information for the Air Force Safety Office to review and digest.

After approximately six months, General Kelly received a message from the Air Force Safety Office, which simply said, "We concur with the San Antonio Position as to the cause of the C-5A accident in Vietnam on 4 April 1975." General Kelly called me at my desk and read the message to me. The weight of the world was lifted off my shoulders. I knew that I now would be able to complete my Air Force career.

I did feel sorry for the 2-Star Board President. He was given one week to retire and lost a star as well. He retired as a 1-Star General, which is not bad. I did feel sorry for him because I think deep down he did agree with our version as the cause but thought he could sell his version in order to deflect the blame from MAC. In most cases, Generals do get their way. The Air Force Safety Office, however, does not play politics especially when loss of life is involved. Their goal is to make flight operations as safe as possible and to prevent similar accidents from happening in the future.

With the Air Force Safety Office's ruling, it was both good and bad for me. As I said, it was good in that I now had the opportunity to finish my Air Force career. It was bad in that now all of the lawsuits that had been filed against the Air Force as a result of this accident now rested solely on me as the Air Force's only authority representative. Russ had retired, and now I was the only one left to defend the Air Force.

For the next 10 years, each time one of these lawsuits came up, I had to go to Washington, D.C., to give a deposition on behalf of the Air Force. This occurred while I was at the Air Command and Staff College, after I left Kelly AFB, and numerous times while I was stationed at Cannon AFB flying F-111s. At least while I was at Cannon AFB, I could take an F-111 to Washington, D.C., and return. This kept my time away from my primary duty to a minimum; however, I had to keep a steel file cabinet full of my accident investigation efforts at home.

Being the Lead Structural Engineer at the time of the accident, I was also the engineer responsible for developing a sure-fire method of insuring that all of the locks did obtain the 3° to 7° over-center position during the locking activity. Essentially, we modified the C-5A ramp locking system by requiring large steel safety pins to be installed in each bell crank after locking. This "Murphy Proofed" the system. As I said, the gang-locking system was adequate but not the best.

The C-5A ramp locking system is a very complex system to properly maintain. The steel pins removed all doubt as to whether the system was properly rigged. The Air Force has never had another C-5A accident as a result of the modified aft ramp locking system.

Note: A Canadian film company, NextFilm Productions, Inc., made a documentary film of this C-5A accident for the "National Geographic TV Network's series "Air Disasters" called "Operation Babylift." This can be found on the Internet.

CHAPTER ELEVEN

The Completion of the Long Return

My second trip to Vietnam in 1975 only reinforced my feelings that everything we had fought for during the Vietnam War was for nothing. All of our sacrifices and loss of lives were for nothing. Our government and the general population just did not seem to care. This only solidified my withdrawal and internalization of my feelings about my experiences in Vietnam.

I was now seeing what many say about war. They say that war is just another political tool for politicians to use. This is very sad and really should not be the case. When war is utilized as an attempt to apply pressure on another government to alter their intentions and/or behavior, it will not accomplish that goal if used improperly. When war is conducted in a manner that the goal of winning is not the overall objective, failure is sure to occur, if proper steps are not accomplished to secure the limited goal.

We military types who have studied war know that there are only two ways to win a war. In his book "On War," Carl Von Clausewitz states that the only way to defeat an enemy is either to destroy the enemy's ability to wage war or destroy his will to wage war. We did neither in the way the Vietnam War was prosecuted. This is why, in my opinion, if a war is to be fought, we should not let politicians decide on the goal that is to be achieved. If politicians do decide the goal, then they should listen to their military advisors on just how to achieve the goal. War is not a political tool for those of us who have to fight it. This is especially true for those 58,000-plus soldiers and airmen who lost their lives in the conflict sometimes referred to as the Southwest Asia War Games (See Photo 37).

While at Kelly AFB, I was selected for Major "below the zone" (BTZ) due to my actions in Vietnam. Even this achievement was tarnished somewhat due to the anti-war and anti-military sentiment that I had experienced upon my return. It is hard to be proud when you feel your actions meant nothing to your country.

One good thing did occur as a result of my work as an Engineer at Kelly AFB. I was able to get my registration as a Texas Licensed Professional Engineer (PE). This helped me tremendously after I retired from the Air Force in 1989.

Being selected for Major below the zone came with the opportunity to attend the Air Force's Air Command and Staff College (ACSC) at Maxwell AFB in Montgomery, Alabama. After graduating from ACSC, I was assigned to Cannon AFB in Clovis, New Mexico, to fly F-111D aircraft. I flew F-111Ds for seven years during which time I was an Assistant Operations Officer of the 524th TFS, Chief of Wing Training for the 27th TFW, and Operations Officer and Squadron Commander of the 523rd TFS. As a result of my actions at Cannon AFB, I was promoted to Colonel and selected to attend the Air Force's Air War College (AWC) at Maxwell AFB.

Upon graduation from the AWC in 1986, I was again assigned to Kelly AFB in San Antonio, Texas. At Kelly AFB, I was the System Program Manager for the T-37, T-38, F-102, F-106, OV-10 and O-2A aircraft. From this position I was moved to Chief of Kelly's budget for Material Management (MM). I was responsible for over five billion dollars of budget authority for the Air Force to procure spare parts.

I retired from the Air Force in 1989 still not ready to talk about Vietnam. As it turned out, however, my tour as a FAC in Vietnam, which was a very long year; but, as a FAC in support of ground troops, it was the most rewarding job that I had in my 25 years in the Air Force.

After retiring in 1989, I took a job as Office Manager and Geotechnical Engineer for an Engineering Consulting Firm (Chenn-Northern). In 1990 Chenn-Northern was acquired by a Holding Company in England that had acquired another engineering consulting firm by the name of Southwestern Laboratories. Since both Chenn-Northern and Southwestern Laboratories had offices in San Antonio, the Holding Company combined the offices. I became the Manager of the Environmental Division of this combined office organization.

In 1992, I departed from Southwestern Laboratories (Environmental Division) to start my own Environmental Consulting Firm. The Holding Company was requiring Southwestern Laboratories to use "Enron" accounting practices that I, as a Professional Engineer (P.E.), just could not accept. In 1992, I started my own Environmental Consulting Firm known as STC Environmental Services, Inc. As Chief Executive Officer (CEO), President and Principle Engineer, I ran this company for 18 years before retiring for good in 2010.

In 2009, my wife bought a book for me titled *Back from War* by 1st Lt. Lee Alley. She purchased the book on March 6, 2009. This book contained stories written by soldiers who had returned from Vietnam and who had experienced the same culture shock and anti-military, anti-war sentiment that I had experienced. Reading this book, I realized that I was not alone. I finally was able to begin to talk about my Vietnam experiences. In one of the stories, the individual had done the same thing with his medals that I had done. I was able to retrieve my medals from the bottom drawer of the steel filing cabinet in our bedroom closet and mount them in a shadow box that my oldest son had given me for Christmas many years earlier.

I now have a room full of Air Force memories in pictures and awards mounted on walls in a room dedicated to 25 years of service in the Air Force (See Photos 38, 39, and 40). I finally came home mentally from Vietnam in April 2009. Thirty-seven years is a very long time for a return. I was still, however, having a problem accepting that our political leaders use the military as just another political tool. I continued to try to understand and justify why this is acceptable considering the cost in American lives and resources that occur during a war and long after. I was taught that it takes approximately 100 years for a country to pay off a war.

CHAPTER TWELVE

The Right to Be Wrong

As I TRIED TO understand the logic and justification as to the use of the military as just another political tool, I thought back to WWII and what the American society's attitude was like then. WWII was fought with the express goal of total defeat of the enemy. Our government leaders, both Republicans and Democrats, along with our allies, were strongly behind the war effort of total defeat of both Germany (Hitler) and Japan. In addition, the majority of the U.S. population also was in total agreement and was firmly behind the war effort and supported our leaders. As our young men went off to war, women worked in the factories building war materials for the effort. The American population knew what the consequences were if Hitler and/or the Japanese were not defeated.

Since WWII was fought with the expressed goal of total defeat of the enemy, it was fought to destroy both the enemy's will and his capability to wage war. Carl Von Clausewitz would have been very proud of how WWII was conducted. The cost of WWII, however, resulted in the loss of over 400,000 American soldiers' lives out of the 16 million Americans that served. Worldwide the cost in lives was approximately 60 million people. WWII was the most costly war in the history of the world. The American generation that fought WWII has been called the "Greatest Generation" because of their unselfish efforts and patriotic dedication to our country.

WWII ended in 1945 with the crushing defeat of Hitler's Army and with the dropping of the atomic bombs on Japan. I think that Carl Von Clausewitz would consider nuclear weapons to be the ultimate weapon for war. These weapons will destroy both an enemy's will and his capability to wage war. With development of nuclear weapons, however, the world environment also changed forever.

At the end of WWII, the United States emerged as a "superpower". Even though the United States and Russia were allies during the war, they really were not friends. It was not long until Russia also developed nuclear weapons in 1949. At the end of the war, Joseph Stalin made it quite clear that his vision of the changed world was one of communist control. He took over much of Eastern Europe and formed the Soviet Union all under communist control. The United States and the western allies had their vision of the world as democratic free societies with individual freedoms and democratic forms of governmental control.

The world environment was now such that the two "superpowers" in the world had dramatically opposed visions of what the world should look like. The Soviet Union had the ideology of Communism, which championed a Marxist/Socialist form of a centrally controlled government. On the other side was the United States which was the champion of free markets and democratic free societies as its ideology. The world was again at war, but now a very different war—a war of ideologies.

This war was a war of opposing ideologies as both countries tried to expand their influence in the world. Since both the United States and the Soviet Union possessed nuclear weapons, it became obvious that the use of these types of weapons would result in total destruction of both countries and most likely the world. This forced the leaders of both the United States and the Soviet Union to accept and adopt a policy of "Mutually Assured Destruction" (MAD). This new war became known as the "Cold War" because a "hot war" was unthinkable.

This resulted in the younger generation in the United States—the "Baby Boomers" and our future leaders—growing up under the uncertainty of the possibility of world annihilation due to the use of nuclear weapons. If a conflict between the United States and the Soviet Union resulted, due to a miscalculation on the part of their leaders or our leaders, it could escalate into a nuclear war. This could happen as each country tried to expand its respective views while simultaneously expanding its influence in the world. As a result, this generation became anti-war and anti-military in their thinking due to the possibility of world annihilation from the use of nuclear weapons.

This totally shaped their approach and attitude toward life, and they became self-centered and less concerned about what was good for the country. They adopted the attitude that society should take care of them; and they had the attitude that "if it feels good, do it." The 1960's was when these "Baby Boomers" became old enough to begin expressing their sense of self-interests. They became the "Hippies, Yuppies and Peaceniks" as they became active and began voicing their ideas, which were significantly different than those of the "Greatest Generation."

This helped explain to me the attitude of the long-haired individual who spat in front of me on my first return from Vietnam. It also helped explain the "hippie" type individual who tapped me on the shoulder in class at the University of Denver on the day before the mid-term exam and asked me what the textbook looked like for the course we were taking. It is from this generation that the present leaders of our country came.

Because of nuclear weapons and MAD, the use of the military as a political tool for use by politicians was born. I do not think that Mr. Clausewitz in 1832 could ever have imagined that weapons would be developed that could cause so much destruction and would essentially change the world as we know it.

The idea was to limit the use of military force to a level of only conventional weapons with a limited objective as a goal. The problem for me is that the cost in American lives and resources is very difficult to justify when the goal is not total defeat of the enemy. This is especially true if the limited goal is not obtained or is lost at a later date.

The first test of the use of the military as a political tool was in Korea. In 1950, the Soviet Union was pushing Communist North Korea to take over South Korea in order to expand Communism. The United States, under the umbrella of the United Nations, had as its goal to stop North Korea from taking over the South. The goal was not to totally defeat North Korea because this would have resulted in a confrontation with the Soviet Union. The limited goal was to apply a limited amount of military force to convince North Korea (and the Soviet Union) to change their goal of expanding communism.

Conflicts between the United States and the Soviet Union were now being fought through proxies. This was a way to avoid direct conflict between the United States and the Soviet Union and to avoid the possibility of escalation to a nuclear level for the conflict. The use of nuclear weapons is the only way a "Superpower" can be defeated. I learned this lesson during war-gaming exercises at both the Air Force's Command and Staff College and the Air War College.

The Korean War lasted from 1950 to 1953. This was only five years after the end of WWII. The cost in lives lost was over four million total with 40,000 being American military lives. The limited goal of stopping North Korea was achieved in 1953 when hostilities ceased, and both North and South Korea remained intact.

At the end of the hostilities, however, when the goal is not total defeat of the enemy, all military leaders know that a very important step is required to ensure that the limited goal is not lost. That step requires that some military troops remain in place to act as a tripwire to discourage the enemy from again pursuing their goal at a later date. Even though the enemy may have currently lost its will to pursue its goal, without destroying the enemy's capability to wage war, they could easily regain the will to wage war at a later date if they saw a possibility to finally achieve their goal.

The United States, through the use of the military as a political tool, had been able to change the will of North Korea in achieving its goal of taking over the South, but North Korea's capability to wage war had not been destroyed. A peace agreement between North and South Korea was never signed. Hostilities ceased with no signed peace treaty between the two countries. To this day, the U.S. still has troops stationed in South Korea as a tripwire in order to sustain the limited goal achieved by the military in 1953. This is a continuing cost to the United States in addition to the 40,000 lives lost during the conflict.

It should be noted that our leaders who had utilized the military as a political tool were leaders from WWII. Harry S. Truman was President from 1945 to 1953 while Dwight David Eisenhower was president from 1953 to 1961—one Democrat and one Republican. At that time there was not much difference in the attitudes of either party toward the military during and after WWII. Both parties supported the need for a strong military as a form of deterrence.

In 1959 Fidel Castro and his Guerilla Army were able to overthrow the American backed Dictator, Cuban President General Batista. In 1960 Castro established diplomatic relations with the Soviet Union. This, of course, was a distinct problem for the United States. The U.S. was going to have a Communist country approximately 90 miles from its shores. The time period between 1961 and 1962 was probably the closest time in which the world was on the brink of a nuclear war as a result of a miscalculation by leaders in both the U.S. and Soviet Union.

In 1961 the U.S. attempted the Bay of Pigs invasion of Cuba, which resulted in a complete disaster. Then in 1962, the Soviet Union, seeing a possible opening as a result of this U.S. disaster, attempted to place nuclear capable missiles in Cuba. Under President John F. Kennedy, a Democrat, it became a test of wills to see who was going to blink first. Thankfully, the Soviets blinked first and removed the missiles from Cuba.

This Cuban crisis in the early 1960's was when I believe that the attitudes of the younger generation were changed from that of the "Greatest Generation." That is when the "Hippies, Yuppies and Peaceniks" became very active in their protest against war. This was the incident that created the environment for the development of the liberal, left wing, anti-war, anti-military component of the Democratic Party.

It was also during this time period that the Communists were trying to expand their influence in Laos and South Vietnam. The first American advisors were sent to South Vietnam in 1961. In 1962, American military units were assisting South Vietnamese troops in their battle against the Viet Cong (VC). Then in 1964, the use of the American military as a political tool was again initiated when North Vietnamese PT boats attacked a U.S. destroyer in the Gulf of Tonkin. There is some question as to this actually happening.

President Johnson, a Democrat, ordered U.S. airstrikes carried out against PT boats and other facilities in North Vietnam after a second attack on U.S. ships. Then in February of 1965, President Johnson ordered bombing missions north of the 17th parallel (DMZ) in retaliation for a VC attack on Pleiku in the Central Highlands. By July 1965, President Johnson had authorized the use of U.S. combat forces to support the South Vietnamese forces. The force level of American troops was increased to 125,000 troops.

This was now the second example of the military being used as a political tool by our leaders. The use of the military was not to totally destroy the enemy but to apply enough pressure to only change his will to achieve his goal. This use of the military was similar to the way the military had been used in Korea. The problem was that the attitude of the American society toward the use of the military was changing as the liberal, left-wing component of the Democratic Party became stronger.

The Vietnam conflict lasted from July 1965 to February 1975 when all financial, military and economic aid was cut-off to South Vietnam by our elected leaders. The left-wing liberal progressive component of the Democratic Party had become strong enough to change our approach to South Vietnam. This resulted in a big difference between South Korea and South Vietnam. The major difference was that no U.S troops were left behind in South Vietnam to act as a tripwire as we had done in South Korea.

It is no wonder that North Vietnam understood that the U.S. attitude had changed and they now knew they could obtain their objective. Essentially, in South Vietnam, the U.S. lost over 58,000 American lives and 150 billion dollars in financial aid; and the limited goal of stopping communism from spreading was not obtained. The strongest country in the world no longer had the will to stop communism through the use of military force when used as a political tool.

On April 30, 1975, North Vietnam took over the South. With this being the case, I had a difficult time justifying the use of the military as a political tool when the loss of American lives was involved. I had grown up under the "Greatest Generation" way of thinking.

With the resignation of President Nixon in August 1974 and when Congress turned down President Ford's request for $720 million dollars in 1975 to assist South Vietnam to stop the North from taking over the South, it was obvious that the American attitude toward war had changed. This was especially true for the Democratic Party and its liberal progressive wing of the Party.

In 1976, the United States turned away from the Republican Party toward the liberal left or the Progressive component of the Democratic Party. There was now a distinct difference between the Democratic and the Republican Parties in the United States and their attitude toward war and the military. Jimmy Carter was elected President and took office in 1977. President Carter was the first of a new kind of President from a new kind of Democratic Party. The Democratic Party was changing its view of the military and its use.

This new Democratic Party began focusing on social issues and was far less supportive of a strong military. In fact, the Democratic Party had changed so much that I doubt if John F. Kennedy would have been able to win the Democratic nomination for President in 1976 since Kennedy was part of the "Greatest Generation."

President Kennedy believed in a strong military, and he even cut taxes. With the liberals focusing on social issues not many of the new Democrats were heard expressing comments like President Kennedy's statement on January 1961 when he said, "Ask not what your country can do for you, ask what you can do for your country."

Under President Carter, the military budget was drastically cut to the point that American military combat capability was adversely affected. I witnessed this first hand when I was assigned to fly F-111s. The F-111D aircraft weapon system was a prime example of the cuts in the military budget. When I arrived at Cannon AFB in New Mexico to fly F-111D aircraft in the late 1970's, the drastic cuts in military spending were evident. At that time, the F-111 Wing at Cannon AFB had four squadrons and a total of 72 F-111D aircraft. When one walked down the flight line and looked into the engine intakes of 19 aircraft, one could see daylight out of the exhaust portion of the aircraft. There were 19 holes in these aircraft where engines were supposed to be installed. The F-111D weapons system was so short on spare engine parts that these 19 aircraft did not even have engines installed. This meant that of the four F-111D squadrons at Cannon AFB, the combat capability was such that it would have been impossible to deploy any squadron to any place in the world in any amount of time to conduct combat operations, if called upon to do so.

As a further illustration of the lack of combat capability, when I arrived at Cannon AFB, the training program that I was scheduled to go through was supposed to take six months or 24 weeks. After completion of this training, I was to be combat ready. It took me one year and one week, or 53 weeks to complete this training. The problem was there just were not enough serviceable aircraft to fly to meet all of the flying requirements of the four squadrons. Essentially, it took me over twice as long as it should have to become combat ready due to a lack of serviceable aircraft and spare parts.

After four years of liberal, progressive left wing policies, the American public became discouraged with high interest rates and a bad economy and the so-called "misery index" (a combination of high interest and inflation) of approximately 18 percent. In 1981, Ronald Reagan, a Republican from the "Greatest Generation," was elected President. President Reagan believed in a strong military and immediately began to rebuild the military back to a level needed to deter the Soviet Union from expanding their influence in the world. A strong military makes a safer world for all.

Another example of why I think that President Kennedy could not get the new Democratic Party's presidential nomination was again evident in his January 20, 1961, Inaugural Address. When talking about our adversaries, he said "We dare not tempt them with weakness. For only when our arms are sufficient beyond doubt can we be certain beyond doubt that they will never be employed." Have you heard any current left-wing liberal Democrat voice a similar statement?

When I left Cannon AFB in 1985, Cannon was down to 68 aircraft and had three flying squadrons. One squadron was a training squadron, and the other two were fully combat ready squadrons. Because of President Reagan, Cannon AFB's combat capability was such that we could now deploy two F-111D combat squadrons to any place in the world within 24 hours of notification and carry on combat operations for as long as needed. This was verified in 1984 when Cannon AFB received its Operational Readiness Inspection (ORI). I was the Squadron Commander of the 523TFS, one of the two combat ready squadrons. The ORI started on a Sunday at 0900 hours (9:00 A.M.).

The way these ORIs are conducted is that an Inspection Team from the Headquarters of the Tactical Air Command (TAC) arrives on base unannounced and starts the inspection. These inspections are usually started at a time that is not a normal time for flight operations. The point is to test to see if the Wing is truly combat ready. It is felt that most wars probably would not start at a convenient time.

Our ORI started at 0900 hours (9:00 A.M.) on a Sunday morning. Once started, we had one hour to contact all of our people and get them back on base to get ready to deploy. In 1984, without cell phones or pagers, this was not always an easy task.

I remember when I received the call that the ORI had started, my Operations Officer, all four of my Flight Commanders and I were scrambling to get our aircrews back on base. I remember going into churches looking for my aircrews in order not to fail the ORI by not being able to get everyone back within the time allowed. This recall not only applied to the flight crews but to all maintenance and combat support people who were involved with combat activities of the Wing.

After the recall, we had to pack up the Squadron while the Maintenance people prepared the aircraft for deployment. Our next deadline for the Wing was to have both Squadrons, all 48 aircraft, ready for takeoff 24 hours after the ORI started.

To ensure that all 48 aircraft of the two Squadrons were actually ready for flight, we had to take off and fly a mission. After the deployment of all 48 aircraft, each squadron was then required to provide 45 combat sorties a day on their 24 aircraft. This was required for three days after which the sortie rate was decreased to 30 sorties a day for each of the 24 aircraft. This was not an easy accomplishment for the complex F-111D aircraft.

The 27th TFW received an overall excellent rating from the ORI exercise as did both Fighter Squadrons. The Wing's total sortie effectiveness rating was 77 percent. The 522nd TFS had a sortie effectiveness rate of 71 percent while my Squadron, the 523rd TFS, had a sortie effectiveness rate of 82 percent. This ORI confirmed that all F-111 models were combat ready. My Squadron of F-111D aircraft was the Replacement Training Squadron (RTU) for the F-111F model that was stationed at RAF Lakenheath, England. After the ORI, a number of my crews were reassigned to RAF Lakenheath, England. On April 5, 1986, a terrorist attack of the La Belle discotheque in West Berlin occurred. During this terrorist attack, three individuals were killed, and 229 innocent people were injured. Two of the individuals killed were American servicemen. Colonel Kaddafi was identified as the person who had sponsored this terrorist attack. President Reagan ordered a punitive attack on Colonel Kaddafi in response to this attack.

Once again the military was being used as a political tool. Just this simple attack had a significant cost associated with it. There were 128 different aircraft utilized to complete the attack. Of the 24 F-111F aircraft that departed England only 23 returned. The F-111 crew we lost was one of my aircrews. We lost two American airmen when the F-111 was shot down and crashed in the Gulf of Sidra. This limited use of the military as a political tool, however, proved successful as Colonel Kaddafi did reduce his role in sponsoring terrorism.

President Reagan was responsible for returning the American military to a position of strength. He was also responsible for defeating the Soviet Union in the "Cold War." He essentially did this through economic policies. He challenged the Soviet Union by indicating that he was going to develop a weapon system that would defeat inter-continental ballistic missiles. He called the system "Star Wars." The Soviets attempted to develop counter measures which essentially drove them into bankruptcy.

On June 12, 1987, President Reagan challenged the Soviet Union's General Secretary Gorbachev while giving a speech at the Berlin Wall. He challenged Secretary Gorbachev to "…tear down this wall!" The challenge was successful and the "Cold War" was won.

After this action, the spread of communism through the use of force was not as much a problem as it had been when the Soviets were in a better position of good economic health. They had to change their approach to expand their influence since they now had to concentrate on taking care of their own population.

When it seemed that the world would be a safer place, however, a new threat was beginning to surface in the form of terrorist attacks. Religious fanatics on many different fronts and locations were conducting these attacks. This included Muslims attacking Muslims as well as attacks on all forms of governments. The attacks had their basis in a fanatic religious ideology as opposed to a type of government control that was in place. This was a new and a very different kind of enemy.

The next use of the U.S. military as a political tool occurred when President George H.W. Bush used the military to stop Saddam Hussein from taking over Kuwait. This became known as the Gulf War. I have also seen it called the Persian Gulf War, the First Gulf War, Gulf War I, the Kuwait War or the First Iraq War. Anyway, it lasted from August 2, 1990, to February 28, 1991. It cost the U.S. $60 billion dollars and 148 U.S. soldiers killed in action.

President Bush was able to put together a Coalition of forces from 13 different countries. This included Muslim as well as Democratic countries. It was the largest military alliance since WWII. President Bush had been in WWII and knew what was required to lead such an alliance of forces. President Bush and the U.S. led this effort and did not lead from behind. President George H.W. Bush was the last President who had the background experience of having been in WWII.

The use of the military as a political tool had as its limited goal of stopping Saddam Hussein from taking over Kuwait. The goal was not to overthrow Saddam's regime in Iraq. Sufficient force was utilized, and the operation was relatively short lived. The operation was successful with very few American lives lost. If there is a proper way to use the military as a political tool, this was it. History proved, however, that it may have been better if President Bush had not stopped, but actually removed Saddam Hussein from the world stage at that time.

Then on September 11, 2001, the world dramatically changed for the U.S. with the attack of the Twin Towers of the World Trade Center and the Pentagon. This attack was not completed by a specific government against the United States but by a group of radical Islamic terrorists using high jacked civilian airliners. This terrorist action was based in a radical Islamic belief and ideology that transcends all borders. The terrorists' goal is to convert the world to their form of Radical Islamic belief under Sharia law and their interpretation of the Qur'an. Islam is not the enemy, but Radical Islam is.

President George W. Bush, the son of the first President Bush, was now the President. He had no choice but to go after the terrorists responsible for the attacks on the Twin Towers and the Pentagon. A fourth aircraft was stopped from getting to its target, but all souls aboard this aircraft were lost when it crashed in Pennsylvania. Once Osama bin Laden had been identified as the leader behind the attack, the use of the military was once again needed.

Since terrorist groups are not specific governments, they will live and exist wherever they can. This makes it very difficult to confront them directly. They are free to move from country to country, and they find it relatively easy to recruit fighters for their cause. This is especially true in the poor underdeveloped countries in the Middle East.

To address this new threat, President Bush put together another Coalition of forces in 2001 to fight terrorism wherever it was found. Because evidence that many of the terrorists were known to be in Afghanistan, U.S. military forces were deployed to Afghanistan to fight terrorists groups like the Taliban and al Qaeda. This was the start of the Afghan War.

Then in 2003, President Bush initiated an invasion of Iraq to go after Saddam Hussein under the pretense that Saddam Hussein had weapons of mass destruction (WMDs). Many people, especially on the left, felt that President Bush made a huge mistake going into Iraq in 2003 in what became known as the Second Gulf War. He did, however, make the decision, and now the U.S. had military forces involved in conflicts in both Afghanistan and Iraq.

This President Bush had grown up under the influence of his father who was part of the "Greatest Generation." This President Bush also was an F-102 pilot in the National Guard and understood the military and its proper use.

In the 2008 Presidential Election, Barack Obama, who was born in 1961 and was raised as a Muslim as a child, entered the presidential race. Mr. Obama had never been in the military and had worked as a Community Organizer before entering politics. His political orientation was far left-wing liberal, and he focused his campaign on ending the wars in both Afghanistan and especially Iraq. He promised to bring our troops home and to "fundamentally change America."

123

With no military experience and being a far left-wing liberal, Obama's basic frame of reference was one of anti-war and anti-military. His election as President resulted in him being in a position to greatly affect the world environment as President of the United States.

By 2011, Saddam Hussein had been removed, and Iraq was a relatively stable country. President Obama, however, being against the Iraq War from the beginning, made the decision to pull all of our troops out of Iraq. Not having any military experience and with his attitude toward the military, he refused to listen to his Military Advisors who recommended that some military troops be left in Iraq to act as a tripwire as was done in Korea.

At the end of 2011, President Obama returned all U.S. troops from Iraq to the States. President Obama justified this by stating that the Taliban and al Qaeda were on the run and no longer represented a problem for Iraq or us.

What was not taken into account by both President Bush and President Obama was the historical discourse between Sunni and Shiite Muslims in Iraq. When Saddam Hussein was in power, he maintained stability through the use of dictorial power and fear. Without this type of presence, no real consensus between the Sunni and Shiite leaders could be established to govern Iraq.

When Syria became a problem, President Obama refused to address the terrorist threat present there or to address the problem represented by the current Syria regime. He justified not addressing either problem by saying the conflict in Syria was a civil war and did not represent a threat to us. He claimed the terrorist activities being conducted were by a J.V. (Junior Varsity) group and, therefore, were not a threat to us. I think President Obama made these decisions based on his Muslim childhood education, his negative attitude toward the military and his strong desire not to be responsible for another war in the Middle East.

The decision to remove all U.S. troops from Iraq and not address the terrorist threat in Syria had resulted in an even more serious threat to the region and the United States. With no U.S. troops remaining in Iraq, there was no tripwire to discourage any group from attempting to take over Iraq when a weak government was fighting among themselves. In addition, not addressing the so called J.V. terrorists group in Syria, the result was the establishment of the very brutal and ruthless Radical Islamic group called ISIS (The Islamic State of Iraq and Syria). This Radical Islamic group has now taken over a large portion of both Syria and Iraq, and now represents a real threat to Iraq and the entire Middle East and the world.

Essentially, with the total U.S. withdrawal from Iraq and no will to address the terrorists in Syria, we essentially had another South Vietnam situation. The U.S. turned its back on South Vietnam, and now we have done the same with Iraq.

After February 1975 when all aid and support was cutoff for South Vietnam, it took the North only until April 30, 1975, to take over the South. We now have ISIS threatening the world and especially Israel.

President Obama also pulled all of our troops out of Afghanistan at the end of 2014. At least for a time, however, he is leaving a small residual U.S. presence with approximately 10,000 troops remaining in-country. It has been recently reported, however, that President Obama plans to remove all troops from Afghanistan by the end of 2016. In addition, President Obama has reluctantly sent a very limited number of advisors and training personnel back to Iraq to assist the Iraqis in fighting ISIS. He also claims that he has built a Coalition to fight ISIS. He is doing this supposedly by leading from behind. All military leaders know that you cannot lead from behind.

In Iraq the U.S. lost a total of 4,491 soldiers between 2003 and 2015. In Afghanistan, the U.S. has lost 2,358 soldiers since 2001. This does not include the significant cost of all the financial aid and support, which is in the billions of dollars for each country. The real question is just what do we have to show for this cost to the United States for all of our sacrifices?

Iraq is about to be taken over by ISIS. We do not know how the Afghanistan situation will play for a while, but the terrorists are again on the move. Syria is the home base for ISIS, and they are on the move in Libya, Yemen and other places. I think the world becomes a very dangerous place when the U.S. is not strong.

I had mentioned that I was taught that it takes 100 years to pay off a war. This is based on taking care of all of the veterans until they pass away. In Vietnam we had over 300,000 wounded, many with permanent disabilities. In Afghanistan and Iraq we had 52,000 wounded soldiers, many with loss of arms and legs due to the improvised explosive devices (IEDs) that the terrorists use so freely. The U.S. currently has four million living disabled veterans from all wars who must be taken care of.

If it does take 100 years to pay off a war, then we still have 30 years to pay off WWII and 50 years for the Vietnam War. We will be paying off the Iraq and Afghanistan War until 2111 and 2114. This, along with the current 18 trillion in national debt, is why I have a hard time justifying the use of the military as just another political tool.

After my experiences in Vietnam and observing how our country has changed since then, I have learned a very important lesson about when the military is used as just another political tool. That lesson is that the military does win conflicts, but the politicians lose the war. They do this in spite of the costs involved.

The only way I have been able to justify the costs associated with the use of military as a political tool is by the following concept. That concept is that the "American society has the right to be wrong." I have to accept this, as it is the only justification that I can come up with. I am, however, glad that I am the age that I am. In addition, the "right to be wrong" does have consequences. Unfortunately, we all have to live with these consequences.

Note: The time period after President Reagan was relatively quiet for the United States. The use of the military during the Clinton Administration (1993 – 2001) was very limited. President Clinton did launch a couple of Tomahawk Cruise missiles at a warehouse on a weekend, but not much more. He did, however, have a significant impact on the military. During his administration, the military underwent significant cuts in combat capability. He ended the Reagan buildup of forces with significant cuts. He justified the cuts as a peace dividend as a result of winning the "Cold War." His cuts were so large that if a country had what he had cut from the military, they would be considered a "super power". Part of the cuts included putting all four F-111 Wings of aircraft in the bone yard. The F-111 fleet represented the total true night, all weather capability for the United States. The F-15E does represent some night, below weather capability but not true all weather capability.

APPENDIX I

Photos

Photo 1: A view of the City of Hue. Note the South China Sea at the top of the photograph. Most of the population in MRI and MRII was located along the east coast of the country.

Photo 2: Hue Phu Bai was the Army base that the 2ⁿᵈ Brigade of the 101ˢᵗ Airborne Division operated out of next to the City of Hue.

Photo 3: This was my living quarters at Phu Bai.

Photo 4: These were our latrines at Phu Bai.

Photo 5: Wooden plywood sidewalks or PSP panels were utilized to get to the shower facility.

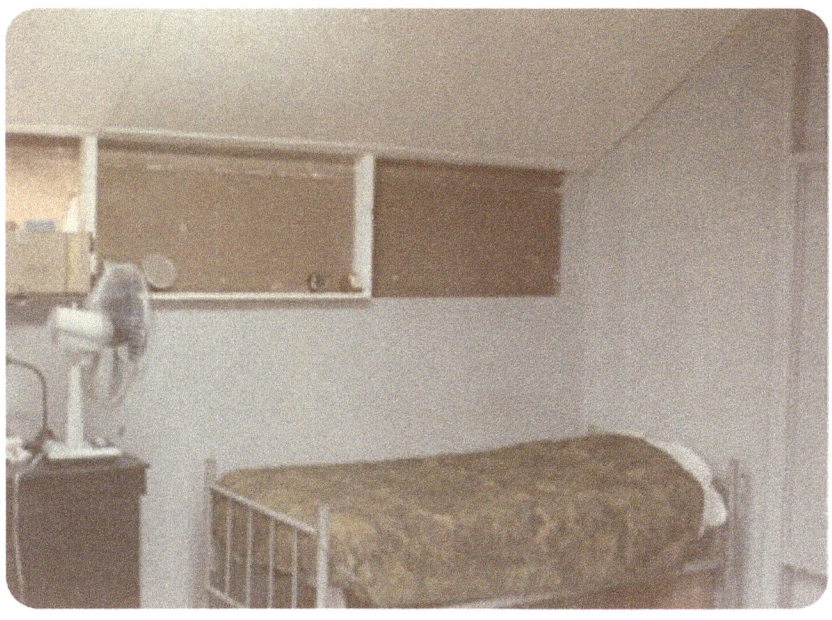

Photo 6: A typical bed inside our living quarters.

Photo 7: A typical steel locker and fan located inside our living quarters.

Photo 8: A typical desk, book shelves, and old rocket boxes being used as cabinets in our living quarters.

Photo 9: View of a typical hooch with a sandbag bunker for use during rocket or mortar attacks. (Bunker is on right side of this hooch.)

Photo 10: View of Post Exchange (PX) at Phu Bai.

Photo 11: View of Fire Base (FB) Rifle in I Corps, August 1971.
This is a typical FB.

Photo 12: View of Fire Base (FB) Nuts in I Corps, August 1971.

Photo 13: View of Interdiction Point "Echo 4" next to Tiger Mountain.

Photo 14: View of Interdiction Point "Echo 7" in August 1971.

Photo 15: This is a photo of me getting ready to fly a mission.

Photo 16: This is a photo of "Little Bird" used as bait to troll for targets during search and destroy missions.

Photo 17: This is a view of the A Shau Valley west of Phu Bai next to the Laos border.

Photo 18: View of the southern end of the A Shau Valley.

Photo 19: View of Laos looking west from the A Shau Valley.

Photo 20: View looking east towards Phu Bai from the A Shau Valley.

Photo 21: View of the Rung Rung Valley southwest of Phu Bai.

Photo 22: View of a typical aircraft parking revetment at Phu Bai.
Note the perforated steel panels (PSP) for the parking surface.

Photo 23: View of a typical aircraft parking revetment at Nha Trang.
Note the parking surface in concrete.

Photo 24: View of the Phan Rang Air Base. Phan Rang was very
much like a stateside Air Force Base.

Photo 25: View of a typical aircraft parking area at Phan Rang.

Photo 26: View of Phan Rang flight line where F-100 aircraft were once sheltered.

Photo 27: The circled facility was my living quarters at Phan Rang. It was the old F-100 fighter squadron's dorm area.

Photo 28: View of the front entrance to our living quarters at Phan Rang.

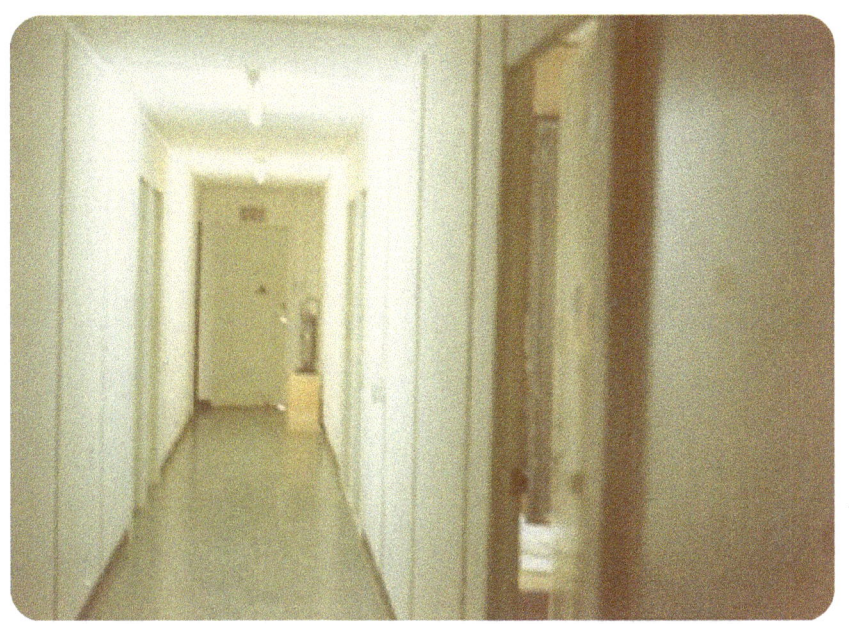

Photo 29: View of the interior hallway of the dormitory type living quarters at Phan Rang.

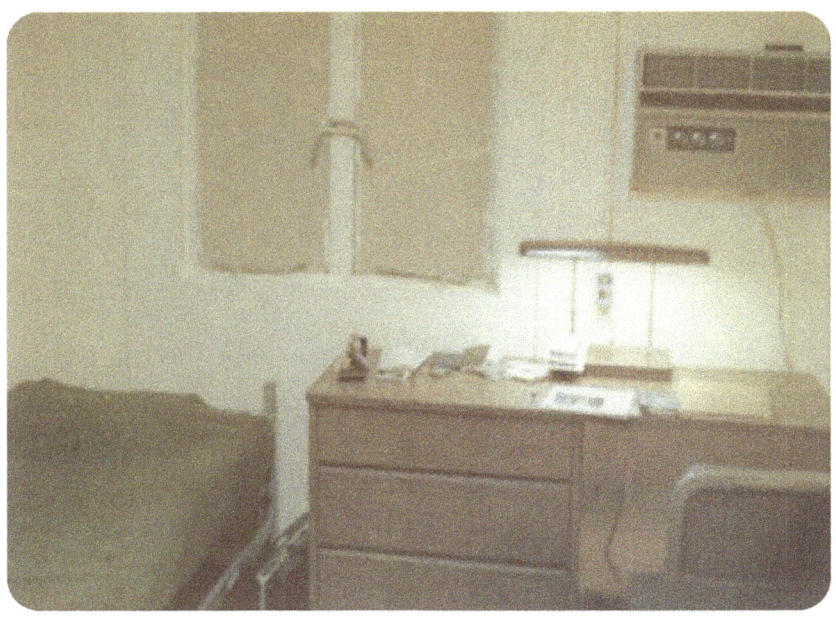

Photo 30: View of the inside of my room at Phan Rang.

Photo 31: Additional view of the inside of my room at Phan Rang.

Photo 32: View of a small hill on Phan Rang Air Base with a road leading to the top of this hill where radio towers were located.

Photo 33: The road on this hill was named McGovern Drive after my neighbor and fellow Instructor Pilot (IP) from Laughlin AFB who had been killed flying F-100 aircraft out of Phan Rang...

Photo 34: View of the parking area for our aircraft at Tan Son Nhut Air Base...

Photo 35: View of the results of a B-52 Arc Lite bombing mission conducted in the A Shau Valley.

Photo 36: This was a sign at Base Ops at Cam Ranh Bay. Home was a long way away from Vietnam.

Photo 37: This is a photo of a patch made to depict that we were participants in the Southeast Asia War Games.

Photo 38: View of my medals mounted in a shadow box that my older son gave me.

Photo 39: View of some of the memorabilia mounted in a room at home.

Photo 40: View of the Vietnam section of memorabilia mounted in a room at home.

APPENDIX II

Figures

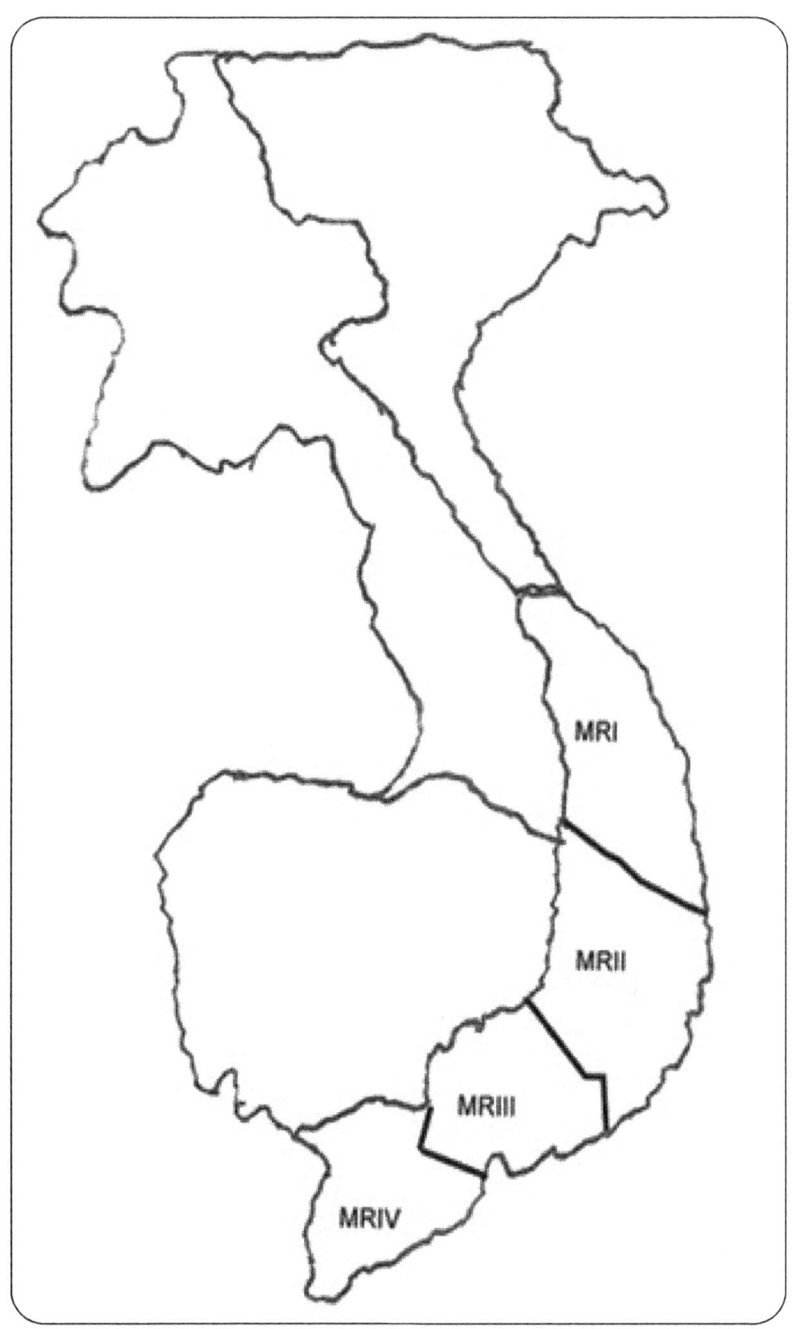

MRI

MRII

MRIII

MRIV

Figure 1: The Four Military Regions of South Vietnam

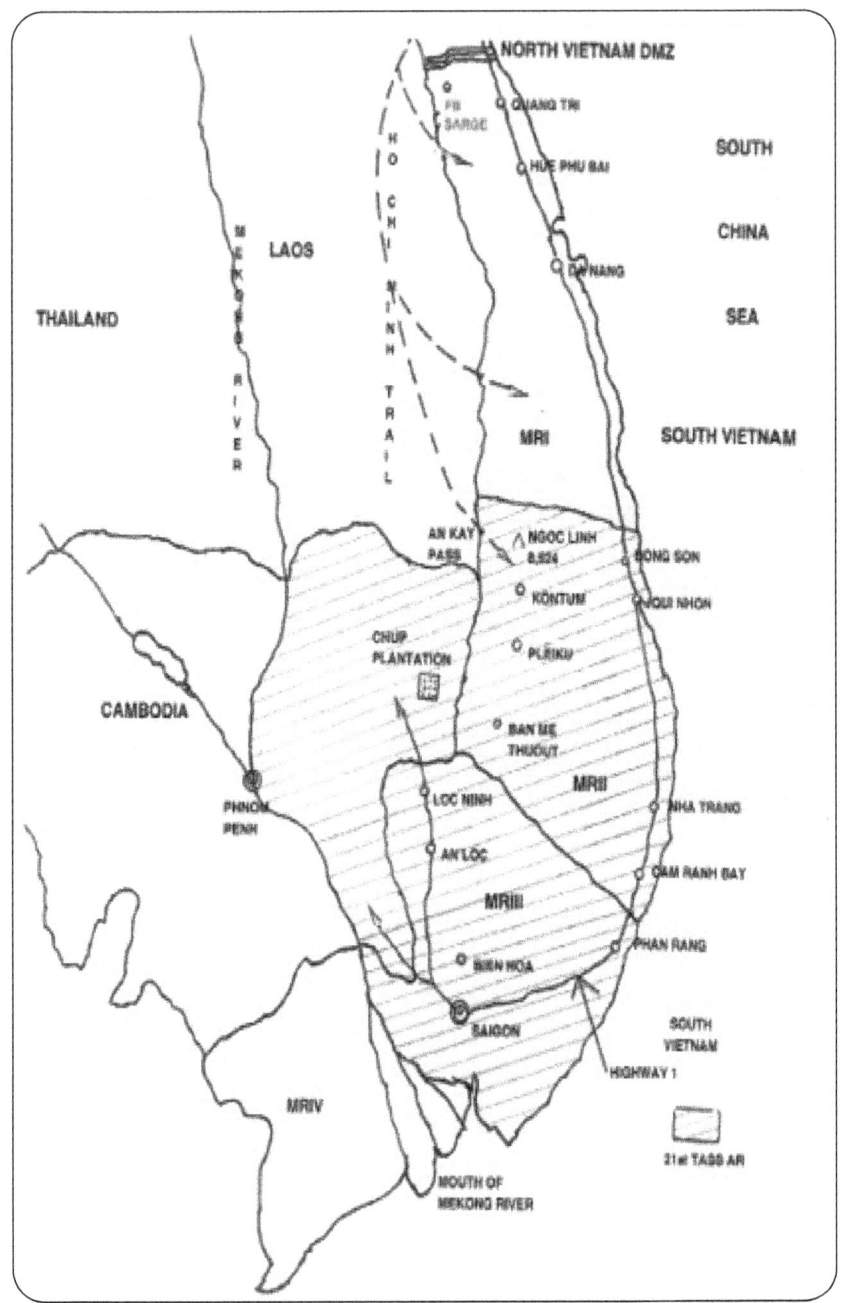

Figure 2: Map showing Base locations and other key points of interest

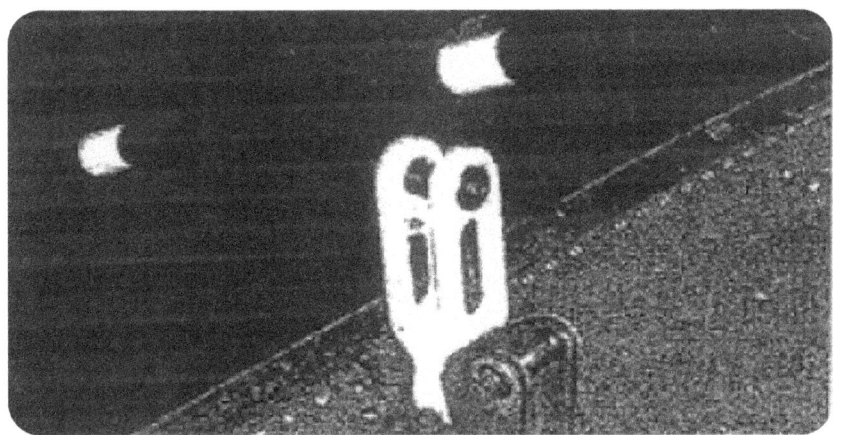

Figure 3: View of Ramp Mounted Stirrup (Lock Clevis).

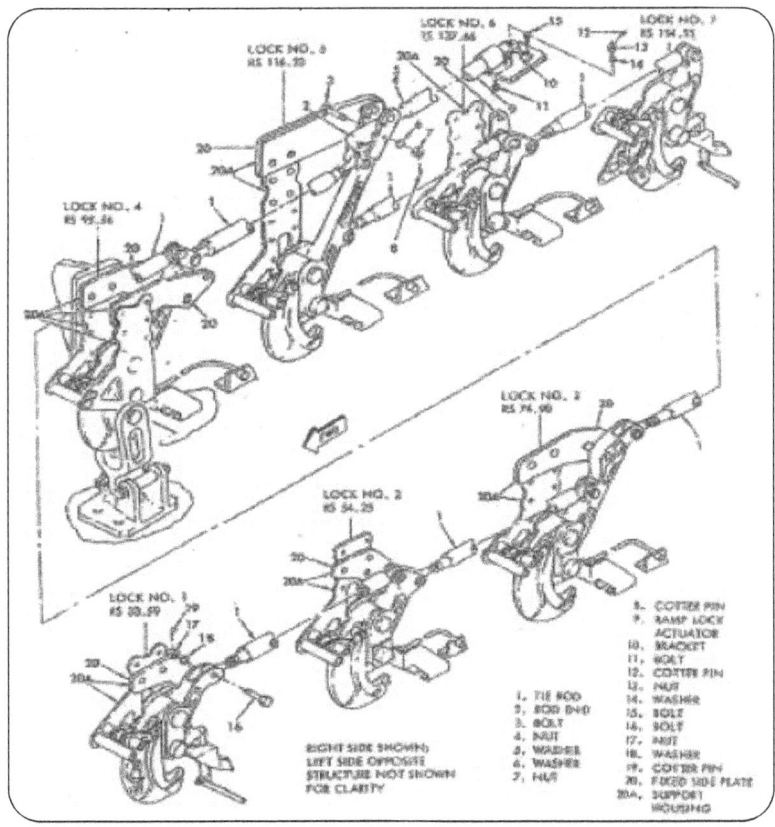

Figure 4: View of Aft Ramp Bellcranks and the Rods

Figure 5: The front of the "Blood Chit"

Figure 6: The back of the "Blood Chit"

We carried the above "Blood Chits" in case we were shot down. The theory was that if we were rescued by locals in the area, if we gave them these, "Blood Chits", they would turn us over to friendlies. I do not know if they worked or not as we never got any FACs back after they were shot down while I was in Vietnam.

POSTSCRIPT

THERE WERE THREE VERY distinct incidents that occurred with the Vietnam War that I cannot forget or forgive. I also know that many others of us who are Vietnam veterans also feel the same way.

The first incident was when Jane Fonda visited North Vietnam and was photographed sitting on an anti-aircraft gun. She did this while we still had American Prisoners of War (POWs) in the "Hanoi Hilton" POW camp. This image just cannot be erased from the minds of us who served in Vietnam and who still had friends who were POWs. This type of anti-war actions and attitudes toward the military and its members caused a tremendous impact on those of us returning from Vietnam. This impact contributed to the distrust, shame and a very unhealthy inward withdrawal by those of us returning home after merely doing our duty as ordered by our government leaders.

The second incident that had a significant impact upon my perspective is when John Kerry, a Vietnam veteran, returned from the Vietnam War and protested against the war. He seemed to be proud when he threw his Vietnam medals over the White House fence in protest of the Vietnam War. He is now our Secretary of State. This left no doubt in my mind what he thought of the medals he had received.

The third incident that caused a significant amount of distrust and disbelief occurred during a CBS Documentary titled "The Ten Thousand Day War." Peter Arnett was interviewing Secretary of State Dean Rusk, and he asked Secretary Rusk if it was true that the United States was providing the North Vietnamese with our next day's bombing targets. Secretary Rusk's answer of "Yes" was totally unbelievable and incomprehensible. To think that our leaders would even consider such a thing, let alone do it, was unthinkable. He tried to justify it on humanitarian grounds. No wonder F-105 pilots flying over the North had approximately 40 percent to 50 percent probability of being shot down—and the Democrats wonder why the military does not trust them! I wonder just how many of my friends were either killed or became POWs as a result of actions like this by our leaders.

To me this was a prime example of a treasonous act by the Secretary of State. It seems to me that if someone is providing our enemy with our planned bombing targets, this is by definition "giving aid and comfort to the enemy." At the least, this was a betrayal or "breach of allegiance" by Secretary Rusk to our Air Force pilots.

It is now 2015, and I just heard that Jane Fonda apologized for the photo on that anti-aircraft gun when she and her husband were visiting North Vietnam. For myself and many of us Vietnam Veterans, there are some actions that just cannot be forgiven. Actions do have consequences! I do not hold grudges, but I do have a very long memory.

In addition, a recent poll indicated that only 15.0 percent of military personnel support the current Democratic administration. When a President agrees to release five Taliban terrorists leaders in exchange for a suspected Army deserter (Bowe Bergdahl), and with at least six other soldiers losing their lives searching for him, along with actions like those of Secretary Rusk, it should not be a mystery as to why.

EPILOGUE

As I LOOK BACK over the past and the time immediately before Vietnam and during Vietnam, I know that there is a higher power at work in this universe. He does have a plan for all of us, and to fulfill his plan, all we have to do is allow him to guide our lives.

God had to be guiding my life as evidence by the timing of the assignment that I received as a FAC. I spent two and half years on the volunteer list before I was selected to go to Vietnam in December 1970. As it worked out, the first 19 pilots selected from the volunteer list at Laughlin AFB did not return from Vietnam. They were either killed in action (KIA) or shot down over North Vietnam and become prisoners of war (POWs).

It was just before I had left Laughlin AFB in January 1971 to go to fighter qualification training at Cannon AFB that Laughlin got three fighter pilots back from Vietnam. If I had not been on the volunteer list for as long as I was, I would have been one of those first 19 pilots who left. God was watching over me and was guiding my life.

A second incident which confirmed to me that God was watching over me occurred the night another pilot and I lost an engine in Military Region II (MRII). This was in the area where "Charlie" did not take prisoners. It was at night, above a cloud deck and in an area with mountains ranging from 3000 feet to 8000 feet. We lost the rear engine, which meant that we were only going to be able to maintain approximately 1100 feet for sustained level flight.

We really could not bail out and could only hope we would make it to the coast over water and then land at Nha Trang AB. God was with us that night as we broke out over water and about three to four miles from mountains that ranged up to 2000 feet. I do not know how the pilot with me felt, but I know God was with us.

I experienced one additional incident where I felt that a higher power was guiding my actions. This occurred on a mission that I was flying out of Tan Son Nhut (Saigon). It was a day reconnaissance mission, and was taxing out to the runway for takeoff. As I approached the end of the runway, I felt a very strong feeling come over me that something was wrong. I had never felt that feeling before on any of my missions, and I never experienced that feeling again during the remainder of my Air Force career as a pilot. A voice in my head said "Don't fly today". I taxied a little farther and the voice got louder and said "Don't Fly Today". I went a little farther and the voice in my head said in a very loud and firm voice "DON'T FLY TODAY". I listened and did not fly that day.

I never had that feeling on any of the other 317 combat missions and the 630 hours of combat flight time that I flew while in Vietnam. I know that God was guiding me that day and telling me not to fly. God does have a plan for all of us; all we have to do is listen, obey him, and let him guide us as he does his work. He also can guide our country, if we let him. Our country was founded on Christian Judea principals. Being a Christian, I believe that God did bless our country and allowed our country to thrive and prosper to the point where we became the greatest nation and influence in the world.

The United States was considered a Christian nation through the time period of the "Greatest Generation." Our actions and laws followed God's teachings and we were blessed. I now have concern as I view the direction in which the liberal progressive left wing of the Democratic Party is taking our country. When President Obama stated that the United States was no longer a Christian nation, I became deeply concerned.

I believe the path we appear to be on was started when the "Baby Boomers" became a generation of inward looking individuals who were more concerned about self-interests and less concerned about the good of the country when the "if it feels good, do it" philosophy became somewhat of the norm for this generation. As I have noted before, this is when I feel the left wing liberal, progressive component was born and began to take over the Democratic Party. This is not to say that only liberals are Democrats because I believe many Republicans have liberal ideas about many social issues as well.

I do believe, however, that this is when the decline in the United States began. I hear politicians keep saying the best days for the United States are ahead of us. I do not believe it. I grew up in the 1950's. As a youth, I can remember that we never had to lock our doors at night. I also had a 1955 red Ford convertible. I could leave the top down and the keys in the ignition at night and never had to worry about it being stolen. People then respected each other and each other's property. Neighbors helped each other and had respect and integrity in their relationship with others.

In the 1960's, when the "Baby Boomers" were growing up under the threat of nuclear weapons and the possibility of world annihilation, the environment changed. Drug use became a major factor in this generation's approach to life and self-satisfaction. Their ideas about sex changed, and a moral decline began to occur. The importance of the family unit was no longer as significant as it had been in the past.

With the increase in drug use, crime rates went up. As the attitude toward sexual behavior changed, the rate of out-of-wedlock births increased. As the "Baby Boomers" grew up under this type of environment, they began to develop their agenda through the formation of the left-wing liberal approach to government as solutions to these problems. This was the start of the United States turning its back on God and his laws and teachings.

The liberal left-wing politicians began to try to remove God from our schools and government agencies by claiming the need for "separation of Church and State." They pursued policies such as legalization of abortion, claiming gay and lesbian activities are normal relationships and that marriage should no longer be defined as just being between one man and one woman. When a nation or people turn their backs on God's laws and teachings, He will step in and punish them. The Bible clearly illustrates what happens when a nation or a group of people turn away from God. When the Jewish people, God's chosen people, turned their backs on God, he sent them into the wilderness for 40 years. He destroyed the cities of Sodom and Gomorrah due to the wickedness of their inhabitants. If we are no longer a Christian nation, God will no longer bless us as he has in the past.

God created us with a free will to choose which direction we want to go. Evil has been at work in this world since Adam and Eve. The Bible describes this on-going battle between good and evil. The Bible also predicts that the final battle between good and evil will occur in the Middle East. It is obvious that evil forces are at work in the world. I do not think that ISIS can be thought of as anything but evil.

Because of my concern about the changes occurring in the United States and the world, I began to study prophecies of the Bible. The Bible prophecies indicate that the final battle known as Armageddon will be fought between the Islamic and the Jewish people. When a peace treaty between Israel and a Coalition of ten Islamic States, that has been set up by an individual known as the Anti-Christ is broken, God will step in to save his chosen people. To do this, God will send his son Jesus back as the second coming of Christ. The second coming of Jesus will accomplish two things.

The first thing that will be accomplished is that those who believe in God will be saved along with the Jewish people who will now accept that Jesus is the Son of God. The second thing that will happen is that evil will be defeated, and the devil and his forces will be cast into the lake of fire forever.

In my studies of the Bible prophecies, I have yet to see where the United States has any major role in the events of the end times. I now believe that what is now happening is all part of God's plan for the world.

No one knows the actual time when God will step in; but, based on the Bible, I feel we are getting near to the final days. After Jesus' first coming, he directed his disciples to spread the gospel throughout the world. With today's Internet and the worldwide capability of communications, this has been done. Not all have accepted the gospel because as the forces of evil have also been at work the entire time.

If Iran does get a nuclear weapon, which it now looks that it is just a matter of time, I feel that they will use it to spread their belief. This will not be good for Israel. The use of nuclear weapons seems to be in line with the Bible's description of Armageddon.

For me, I no longer worry about what is happening, and I will leave it up to God to complete his plan for his creation. I do think that the end can be delayed, but this country would have to return to God. I am not sure that there are enough Christian people or those who believe in God's teachings to rise to the task. If there are, they will have to get to work soon. I do not feel that it is impossible to change the current United States' path because anything is possible with God's help. After all, David defeated Goliath with God's help. A recent poll, however, indicates that Christianity is on the decline in the United States.

At my age and with the above understanding of God's plan, I will leave it up to him to complete his plan for his creation on his timeline. I am now completely at peace with myself and with what may happen. Our American society does have a right to be wrong, but there will be consequences. Even the Roman Empire found that out.

2025 UPDATE

IT IS NOW 2025 and I truly believe that God has decided to give the United States a second chance. I believe this is evidence by the fact that it had to be God who stepped in during Donald Trump's presidential campaign rally on July 13, 2024, at Butler, Pennsylvania. Only God could have caused Present Trump to turn his head at the precise instant that caused the assassin's bullet to strike his ear instead of his head. This had to be Devine intervention by God. President Trump then went on to win the 2024 election in a landslide with a strong mandate from the American people. He won the popular vote 77,302,580 to 75,017,613 and all seven swing states. God is not finished yet with the United States.

ACRONYMS

ACSC	Air Command and Staff College
ADC	Air Defense Command
AFB	Air Force Base
AFIT	Air Force Institute of Technology
AFLC	Air Force Logistics Command
AFSC	Air Force Systems Command
ALC	Air Logistics Center
ALO	Air Liaison Officer
AR	Area of Responsibility
ARVN	Army of the Republic of Vietnam
ATGSB	Advance Test for Graduate Study for Business
AWC	Air War College
BDA	Bomb Damage Assessment
BS	Bachelor of Science
BTZ	Below the Zone
CC	Command and Control
CEO	Chief Executive Officer
CDIPR	Crash Data Indicator Position Recorder
CIA	Central Intelligence Agency
DASC	Direct Air Support Center
DFC	Distinguished Flying Cross

DMZ	Demilitarized Zone
DNIF	Duty Not Involving Flying
DOS	Date of Separation
FAC	Forward Air Controller
FB	Fire Base
FBI	Federal Bureau of Investigation
FOL	Forward Operating Location
FRAG	Fragmentation Order
IED	Improvised Explosive Device
IFR	Instrument Flight Rules
IP	Instructor Pilot
ISIS	The Islamic State of Iraq and Syria
ISU	Iowa State University
JV	Junior Varsity
KIA	Killed in Action
LRRP	Long Range Reconnaissance Patrols
LT	Lieutenant
LTC	Lieutenant Colonel
LZ	Landing Zone
MAC	Military Aircraft Command
MACV	Military Assistance Command Vietnam
MAD	Mutually Assured Destruction
MM	Material Management
MR	Military Region (MRI, MRII, MRIII, and MRIV)
MS	Master of Science
NVA	North Vietnam Army
ORI	Operational Readiness Inspection
PACAF	Pacific Air Force
PE	Professional Engineer
PSP	Perforated Steel Planks (or Panels)

POW	Prisoner of War
PX	Post Exchange
ROE	Rules of Engagement
ROTC	Reserve Officer Training and Commissioning
SAC	Strategic Air Command
SAT	Scholastic Aptitude Test
SF	Special Forces
SPO	System Program Office
TAC	Tactical Air Command
TACAN	Air Navigational Aid
TASC	Tactical Air Support Center
TASG	Tactical Air Support Group
TASS	Tactical Air Support Squadron
TDY	Temporary Duty
TFS	Tactical Fighter Squadron
TFW	Tactical Fighter Wing
TIC	Troops in Contact
TOC	Tactical Operations Center
US	United States
USAF	United States Air Force
UHF	Ultra High Frequency
UPT	Undergraduate Pilot Training
VC	Viet Cong or "Charlie"
VNAF	Vietnamese Air Force
VNMC	Vietnamese Marine Corp
VFR	Visual Flight Rules
WMD	Weapons of Mass Destruction
WSO	Weapons System Operator

www.ingramcontent.com/pod-product-compliance
Lightning Source LLC
Chambersburg PA
CBHW051152120626
46547CB00012B/1058